For Julie

Thanks very much
for your hospitality

Best wishes,

Ann Barnder

Sept 14, 2003

::

Better Than Prozac

::

Also by Samuel H. Barondes

Cellular Dynamics of the Neuron
Neuronal Recognition
Molecules and Mental Illness
Mood Genes

: :

Better Than Prozac

: :

CREATING THE NEXT GENERATION
OF PSYCHIATRIC DRUGS

Samuel H. Barondes

OXFORD
UNIVERSITY PRESS

2003

OXFORD
UNIVERSITY PRESS

Oxford New York
Auckland Bangkok Buenos Aires Cape Town Chennai
Dar es Salaam Delhi Hong Kong Istanbul Karachi Kolkata
Kuala Lumpur Madrid Melbourne Mexico City Mumbai Nairobi
São Paulo Shanghai Taipei Tokyo Toronto

Copyright © 2003 by Samuel H. Barondes

Published by Oxford University Press, Inc.
198 Madison Avenue, New York, New York 10016

www.oup.com

Oxford is a registered trademark of Oxford University Press

Library of Congress Cataloging-in-Publication Data

Barondes, Samuel H., 1933–
Better than prozac: creating the next generation
of psychiatric drugs/by Samuel H. Barondes.
p. cm. Includes bibliographical references and index.
ISBN 0-19-515130-5
1. Antidepressants—Popular works. I. Title
RM332.B37 2003 616.85'27061—dc21 2002033396

1 3 5 7 9 8 6 4 2
Printed in the United States
on acid-free paper

For Louann

And to welcome
Ellen Ariel

Contents

Writing prescriptions is easy, but understanding people is hard.
—Franz Kafka, *A Country Doctor*

A Note on Drug Names

Medications have two names: a trade name, such as Prozac, which is capitalized and is chosen by the drug manufacturer in the hope that it will be attractive and memorable, and a generic name, such as fluoxetine (Prozac's generic name), which is not capitalized and is chosen by scientists on the basis of the chemical structure of the drug.

To complicate matters, some medications have more than one trade name. For example, Eli Lilly, the manufacturer of Prozac, sells the same drug as Sarafem, a name chosen to make it more attractive to women with premenstrual symptoms. The same drug may also have different trade names in the United States and the United Kingdom. Although Prozac has the same name in both markets, the related drugs called Zoloft and Paxil in the United States are called Lustral and Seroxat in the United Kingdom.

In this book I will use widely known trade names for some medications such as Prozac, and generic names for others. The appendix lists the generic names of the main drugs covered by this book, along with trade names in the United States and the United Kingdom.

Prologue

In the early 1960s, when I began my training in psychiatry, the field was in a state of great excitement. In the previous decade several drugs that had been developed for other purposes were unexpectedly found to be useful for the treatment of mental disorders. For the first time a daily dose of pills could eliminate the bizarre ideas of people with schizophrenia and relieve the agonizing self-hatred of people with severe depression.

Adding to the excitement was a flurry of findings about the effects of these drugs on neurotransmitters, the brain chemicals that transmit signals between nerve cells. Once this became clear, a search was undertaken for still more chemicals that influence the actions of neurotransmitters. This led to the development of many new medications, some after long periods of experimentation. The most famous, fluoxetine (Prozac), was identified in the early 1970s[†] as a chemical that augments neurotransmission by serotonin, and marketed as an antidepressant in 1987. Related drugs soon followed, joining many others as tools for changing neurotransmission in the brain.

These new drugs, which have a better balance of properties than those discovered in the 1950s, transformed the practice of psychiatry.[†] Before they came on the scene, most psychiatrists were mainly interested in psychotherapy and rarely prescribed pills. They believed that drug treatment should be restricted to the small fraction of their patients with severe mental disorders and frowned upon the use of drugs for milder forms of mental distress. Now most peo-

---

[†]Dagger refers to the notes at the end of the book.

ple who consult psychiatrists leave the office with a prescription for at least one medication. More than a hundred million people around the world are presently taking psychiatric drugs. Many more are wondering whether they or their loved ones might also benefit from these medications.

The aim of this book is to describe the origins and properties of these psychiatric drugs and to explain how new ones are being created. Part I is about the evolution of those that are widely prescribed today.[†] The main point I will make is that the latest ones are generally improved versions of those that were accidentally discovered decades ago, and that it will be difficult to make further improvements until we have a better understanding of the origins of mental disorders. The emphasis of Part II is on genetic and neurobiological studies of these disorders and of the effects of psychiatric drugs on the brain. These lines of research are reinvigorating this field and will guide the design of more effective medications.

With regard to my biases, let me confess at the outset that I have been an enthusiastic fan of psychiatric drugs for more than forty years. I have prescribed them for patients with a wide range of ailments and have seen many excellent results. I have worked with them in the laboratory, taught about them to medical students and young psychiatrists, given advice about them to public agencies and private foundations, and participated in the creation of new ones by serving on the scientific advisory boards of pharmaceutical and biotechnology companies.

But I have also seen these drugs sometimes fail miserably because of their many flaws. Even the best of them are blunt instruments that have a large number of effects on the brain, only some of which can be considered therapeutic. As you grow familiar with these fascinating chemicals[†] in the course of this book, you will come to understand the reasons for their limitations and what must be learned before we can expect substantially better ones.

PART I :: SWEET OBLIVIOUS ANTIDOTES

Canst thou not minister to a mind diseas'd,
Pluck from the memory a rooted sorrow,
Raze out the written troubles of the brain,
And with some sweet oblivious antidote
Cleanse the stuff'd bosom of that perilous stuff
Which weighs upon the heart?
—William Shakespeare, *Macbeth*

1 :: *Clara's Secret*

Truth emerges more readily from error than from confusion.
—Francis Bacon (1620)

Every morning, after breakfast, Clara takes a psychiatric drug. She doesn't feel any different after gulping it down with a large glass of water. But Clara is afraid that without this medication she will quickly revert to the self-doubting and depressive state that led her to it in the first place. She has come to believe that, for this and other reasons I will get to, the pill-taking ritual is essential to her continued well-being.

Clara began taking this medication in December 1993, shortly after her twenty-sixth birthday, during a period of great emotional distress. A tall, lean, attractive redhead with a brilliant intellect, she had recently been granted a leave of absence from a graduate philosophy program at a university in northern California, where she had been writing a dissertation on the work of Francis Bacon, the seventeenth-century British philosopher. Uncertain about her long-range goals, Clara had decided to spend a few years teaching English at a primary school for girls in San Francisco. But she soon began to have doubts about stopping her graduate studies, and spent her nights worrying rather than sleeping. A mutual friend, Martin, a philosopher who had been one of Clara's mentors, asked if I would see her because her turbulent self-doubt had a quality of hopelessness that she had not displayed previously. Having tried in the past to convince Clara to stick with her studies, Martin now was worried that her withdrawal might lead to a downward spiral of despair. Two weeks later, in my second meeting with Clara, I gave her a prescription for the pill that she has taken ever since.

::

Pills were not always a form of treatment that Clara had been willing to try. Three years earlier, in 1990, during another period of indecisiveness and depression, Clara had resisted a suggestion from a fellow student that she might benefit from medication. Instead, she found relief in psychotherapy. In the course of sixteen weekly sessions with a psychologist who specialized in a technique called cognitive therapy,[†] Clara gradually began feeling better. Abandoning self-doubt, she became much more optimistic about the career she might have in philosophy.

It was during this period that Clara turned her attention to Francis Bacon's most famous book, *The New Organon*. Published in 1620, when Bacon was almost sixty, it argues that knowledge comes mainly from experiments and observations instead of from simply applying old ideas handed down by learned authorities. Bacon's argument, which we have come to take for granted, was revolutionary in its time. *The New Organon* played a pivotal role in the development of experimental science.

As Clara read more by Bacon, and about him, she came to think of this seventeenth-century gentleman-scholar as a kindred spirit. She was particularly attracted by Bacon's scathing criticism of the human tendency to make inaccurate inferences based on inadequate information. Relating Bacon's message to herself, Clara began to recognize her own propensity to make unfounded and excessively critical self-judgments. When viewed in this way, it seemed to Clara that Bacon's writings were important not only for the overall development of science, but also as a basis for the changing views of human nature that came with the Age of Reason. Pleased with this insight, she decided to write her Ph.D. thesis on the psychological implications of the epistemology of Francis Bacon.

Clara later came to believe that her fascination with Bacon arose in part from her experience with cognitive therapy, a form of treatment that had seemed to be made to order for her. Based on the idea that certain types of mental anguish, especially depression, are due to maladaptive interpretations of life events, the aim of cognitive therapy is to teach clients new ways of thinking that lessen their chances of feeling depressed. In Clara's case, her therapist explained that people who are inclined to be depressed have inaccurate negative views of themselves, the world, and their prospects for the future. She also explained that it doesn't matter if these negative views are mainly reflections of innate temperament or of life experiences. Whatever the origin of this bleak and self-denigrating perspective, the remedy is the same: to learn to think more positively by reassessing situations through conscious mental effort. To Clara, who enjoyed conscious mental effort, this sounded like an appealing prescription.

Armed with the more positive outlook that she acquired through cognitive therapy, Clara had several good years. She immersed herself in the writings of Bacon and became interested in the stories of his personal successes and misfortunes. She was particularly inspired by Bacon's ability to split his time between philosophical writings and important public responsibilities in Parliament. She was fascinated by his rise in 1617 to lord chancellor, the highest-ranking law officer in England, and his dismissal from that position in 1621 for accepting bribes. Bacon became, to her, not only a man of ideas, but also a vivid personality to whom she felt very close, despite his shortcomings.

Then, after finishing an initial draft of her doctoral thesis in a state of high enthusiasm, Clara's self-doubt came back with unprecedented force. When Martin suggested that she return to the cognitive therapist, she told him she had decided that it was a waste of time. Pressing her to try it again, he was shocked to hear Clara dismiss cognitive therapists as "panderers of false optimism, dispensing rosy-colored nonsense." It was while she was in this frame of mind that I first met Clara.

The main reason Martin sent Clara to see me was to find out if she might benefit from an antidepressant drug. He told me that she had become fascinated with these medications but was wary that they might be as disappointing as psychotherapy. He hoped that I would figure out what was troubling Clara and decide on an appropriate remedy.

To do this job, I needed to learn more about Clara's life story and to decide how her pattern of symptoms matched with those described in psychiatry's diagnostic manual, *The Diagnostic and Statistical Manual of Mental Disorders* (abbreviated *DSM*). During a lengthy interview I concluded that her long-standing depression fit best into the category called dysthymic disorder, and that it was not severe enough to be classified as a major depressive disorder—the category of depression that is usually treated with medications. In considering her treatment by the standards of practice of 1993, I was reluctant to prescribe an antidepressant because the risks might be greater than the rewards. Instead I suggested that she go back to the therapist who had helped her in the past.

Clara vigorously rejected this suggestion. She said that she had learned the lessons of cognitive therapy very well and had tried to apply them in her current situation. But they didn't work anymore. "Look," she said, "I personally like my therapist, and she taught me a lot. But more of the same treatment just doesn't make sense. I'd rather take pills to fix whatever is wrong with the chemicals in my brain. But I'm very frightened of pills, because they might turn me into a zombie. So I'm resigned to accepting that I'm just a hopeless mess."

Having heard these arguments, and remembering Clara's comments to Martin about cognitive therapy, I was persuaded that she had given up on that approach. Unwilling to just send her home with no more treatment than our conversation, and recognizing that she had severe insomnia, I prescribed a small bedtime dose of amitriptyline (Elavil), an antidepressant that also happens to be sedating. I made this proposal because a prompt improvement in her sleep might get her on the course to recovery. The main reason

I didn't prescribe a conventional sleeping pill is that they sometimes make depression worse. I also thought it would be useful to see how Clara responded to a drug that has some antidepressant effects, even though those effects appear only after weeks of treatment.

Explaining my rationale to Clara, I told her that I was proposing what is called an off-label (i.e., not officially approved) use[†] of amitriptyline, because the drug is not an established treatment for either insomnia or mild depression. I also reassured her that its sedative action was well known, and that it was acceptable medical practice for me to give her this drug, even though its effectiveness had not been proved in formal clinical trials with patients like her.

When my proposal for a brief trial of amitriptyline was presented in this way, Clara agreed to try the experiment. So I gave her some samples and asked her to take 50 milligrams of amitriptyline—one-third of the typical antidepressant dose—half an hour before bed-time. I also asked her to keep a diary to record the time she took the pill, how long she slept, and how she felt during the day. Unless she started to feel worse, she should come back to see me in a week. If she was becoming a zombie, she should call me right away.

::

The first night Clara took amitriptyline, she slept for nine hours straight. Waking refreshed, she noticed that, as I had warned her, her mouth was dry—a well-known effect of the drug, which reduces the secretion of saliva by blocking the action of a chemical called acetyl-choline. But this was only mildly annoying, and the long period of uninterrupted sleep seemed to propel her through the day. She was unusually alert in the classroom, and the twelve-year-old girls she was teaching seemed to be asking more interesting questions.

The improvement in Clara's sleep persisted through the first week of bedtime medication. Although the effects were not always as dramatic as they were after the initial dose, Clara was very happy with the result. On the fourth night of amitriptyline she had tossed and turned for about an hour after taking the pill, until sleep finally arrived. Nevertheless, she awoke feeling well rested. When she came to my office after a week on this regimen, she showed me a diary full of positive comments. Life with more sleep was clearly better.

But the relief was only partial. Although Clara was eager to reassure me that she was now on the mend, she admitted that she still felt depressed. The normal sleep was not enough to undo a set of problems that she had been grappling with for many years. Now that she was calmer and thinking more clearly, we could talk about these problems in more detail.

She had already told me about her struggles with her mother, Fanny. In Clara's view, Fanny, a schoolteacher, demanded nothing less than perfection from her daughter. And Clara, Fanny's only child, was for many years happy to comply. Then, while in college, Clara started accusing Fanny of being too demanding. Cognitive therapy had made Clara aware of this pattern of behavior and helped bring peace between her and her mother.

Once this peace was achieved, Clara stopped blaming her propensity for depression on Fanny's extreme standards. Instead she came to the conclusion that she was innately different from most people. Knowing that her father had once been so depressed that he had taken a leave of absence from the small bank where he worked as an assistant manager, Clara began to wonder if she had inherited this tendency from him. Perhaps some genes that he had passed on to her had given rise to a defect in her brain that even her well-schooled reasoning abilities could not overcome. It was, in fact, this new attitude about herself that had driven her from graduate school. How could she hope to become a professor of philosophy if she wasn't even capable of trusting her own brain?

Feeling more comfortable after a week of restful nights, Clara acknowledged that she had actually been worried about her brain for quite a long time. She also confessed that she had been fantasizing for several years about being rescued by antidepressant pills. But she had struggled against this inclination, and had been reluctant to admit it when we first met, because she took it as a sign of weakness—an admission of the fallibility of not only her brain but also her character. Having obtained some benefit from amitriptyline, she was ready to express the hope that medications might help her:

"Maybe I do have a chemical imbalance in my brain that can be corrected with pills."

::

The idea of a chemical imbalance in the brain had come to Clara's attention from extensive coverage of this topic in magazines and newspapers. Having read many articles about the behavioral effects of brain chemicals such as dopamine, norepinephrine, and serotonin, Clara was fascinated by the proposal that different balances of these and other chemicals might make people more or less confident or self-critical. The articles also heralded the arrival of new medications that could manipulate brain chemicals with scientific precision, thereby alleviating many forms of mental distress. For example, if someone had a deficiency of brain serotonin, it could be fixed by a medication that raised the level of serotonin to normal. Although Clara recognized that such claims might not be completely accurate, she longed for the relief that they promised—relief she had found so hard to achieve by way of self-reflection and cognitive therapy.

It was, in fact, through the popular press that Clara first laid her eyes on the green-and-ivory pill called Prozac (fluoxetine), a great big zeppelin of a pill floating in welcoming clouds on the cover of a 1990 issue of *Newsweek,* along with the headline "A Breakthrough Drug for Depression." Several years later she gobbled up Peter Kramer's *Listening to Prozac,* which called attention to the widespread practice of prescribing this and other medications for people with levels of mental distress that seemed mild compared to her own. Although not everyone who wrote for the public was in favor of the burgeoning use of these new medications, it was hard to ignore the outpouring of testimonials.

Clara told me that she was particularly struck by *Time* magazine's description of Susan Smith, "a [44-year-old] self-described workaholic [who] needs a little chemical help to be a supermom: she has been taking the antidepressant Prozac for five years." In the October 11, 1993, article from *Time,* which Clara gave me, the author, Anastasia Toufexis, went on to explain: "Smith never had manic-depression or any other severe form of mental illness. But before Prozac, she suffered from sharp mood swings, usually coinciding with her menstrual periods. 'I would become highly emotional and sometimes very angry, and I really wasn't sure why I was angry,' she recalls. Charles [her husband] will never forget the time she threw her wedding ring at him during a spat. Now, says Susan, 'the lows aren't as low as they were. I'm more comfortable with myself.' And she has no qualms about her long-term relationship with a psychoactive pill: 'If there's a drug that makes you feel better, you use it.'"

It was this article in *Time* that had persuaded Clara to think seriously about taking medications. The carefully folded article, which she had pulled from her briefcase, showed a photo of Susan Smith in a position of tranquil repose. Now that Clara had had her taste of psychopharmacology with a successful response to the sedative effect of amitriptyline, she said that she was ready for the next experiment. But an increased dose of amitriptyline was not appealing to her because the very same sedative effect that she was already using to good advantage at bedtime would make her sleepy throughout the day. Following the example of Susan Smith, Clara wanted to find out if she, too, would benefit from that new wonder drug, Prozac.

::

Clara's plea for Prozac put me on the spot. Unlike the off-label use of amitriptyline as a sedative, which was a common practice, using Prozac off-label for patients with moderate depression (dysthymic disorder) was at that time frowned upon by some of my most respected colleagues. They felt strongly that prescriptions for Prozac should be given only to people with major depression—the type of person who was known to benefit from this drug. Clara was not considered to be eligible for Prozac treatment because she didn't have the full lineup of symptoms and the degree of disability that are required to make a diagnosis of major depression.

Yet Clara hated the way she felt. Furthermore, she had been depressed before, suggesting that she might have a lifelong vulnerability to excessive moodiness. In addition, her self-doubt had caused her considerable harm by leading her to abandon the doctoral studies that she found so rewarding. Having tasted the benefits of adequate sleep, she now longed for the greater benefits that might come if she received the same treatment as Susan Smith. After all, she and Susan Smith seemed to have a similar problem—something less intense than major depression, but sufficiently severe and disruptive to justify a trial of any available treatment that might be helpful. If Susan Smith could benefit from an off-label prescription of Prozac, why shouldn't Clara be allowed to try it?

In responding to Clara's request I explained how difficult it would be to interpret the results of the experiment she proposed. What would we conclude if she felt better after taking Prozac? As with many other drugs, there is the problem of distinguishing a positive pharmacological effect (reflecting a drug-induced remedial alteration in body chemistry) from the well-known placebo effect (reflecting the effect of pills that is due to the faith and hope of the patient). How would we decide if any improvement observed while she (or Susan Smith) was taking Prozac was due to a placebo effect or a true therapeutic drug effect?

The way placebo effects are distinguished from true drug effects is through an experimental procedure called a randomized controlled clinical trial or a placebo-controlled double-blind study. In this procedure, people with a particular disorder are randomly assigned to one of two groups. Members of one group are given a pill containing the active drug. Members of the other group receive a look-alike that contains no active drug (the sugar pill, or placebo). The results are evaluated by appropriate tests. In the case of studies of antidepressants, the tests are a series of questions that measure mood. What makes the study double-blind is that neither the patient nor the doctor making the evaluations of mood knows who is getting the active drug and who is getting the placebo. Only the researchers in charge of the study have access to that information.

Before Prozac and amitriptyline became widely accepted as treatments for major depression, double-blind studies were carried out in thousands of patients. The upshot of these studies was that more than half of people with major depression showed significant improvement after several months of treatment with either Prozac or amitriptyline, and that the drugs are more effective than placebos. But the studies also showed that many people who suffer from major depression start feeling better while taking a placebo[†]—an effect of pills that has nothing to do with a drug-induced alteration in brain chemistry.

Knowing of the substantial placebo effect of pills used to treat major depression made me wary of starting Clara on a course of treatment with Prozac. I was particularly reluctant to do this because there was not yet any formal evidence that Prozac was an effective treatment (i.e., better than placebo) for patients with dysthymic disorder. If Clara felt better on Prozac, the seemingly successful experiment might condemn her to years of taking a drug that was really no better than a sugar pill. Was Clara prepared for this type of result?

Her answer was an unequivocal yes. "Let me just feel better," she said, "and we'll worry about it later. And please don't mix up the unproven efficacy problem with the placebo problem."

In effect, what Clara was saying was that even if Prozac were an officially approved treatment for people with dysthymic disorder, there would still be the placebo problem. In all placebo-controlled studies it is not clear which individuals get better because of a placebo effect and which get better because of a drug effect.[†] So there is uncertainty anyway. As Clara—Bacon's student—put it, "The truth of the matter is that with drugs like this, which don't always work, everyone who takes them is really conducting their own personal experiment. I'm willing to give Prozac a try, because I don't have much to lose. If it doesn't work, I'll stop taking it. If it does, or just seems to, we might do some other experiments."

In the face of this argument, I agreed to give Clara a prescription for Prozac. Because there was no established dose for patients like her, I decided to start with 5 milligrams a day, a quarter of the usual dose for major depression. I told her she could continue to take the amitriptyline every night for sleep but should feel free to skip it if she didn't think it was necessary. I also asked her to record a rating of how she felt every day at noon, using a scale of 1 (worst) to 10 (best), as a way of evaluating her progress.

::

In the first week of Prozac there was not much change in Clara's mood. After taking the initial dose she thought she felt better. But it didn't last. Her quality-of-life ratings through this period ranged between 3 and 4—no different from the 3 she had assigned herself on the day I had given her the prescription. She had continued to sleep well and so had been reluctant to give up the amitriptyline. Her overall impression was that the Prozac wasn't doing anything, either good or bad.

On the basis of this report I increased Clara's daily dose to 10 milligrams of Prozac for the next two weeks. Again there were no obvious changes in her demeanor, her ratings, or her reports. The only thing she noticed was that she had begun to feel less concerned than usual about everything ("a little detached, but not uncomfortably so"). She also found that she could now sleep well without amitriptyline and used it only occasionally. None of this changed when I first raised her dose to 20 milligrams.

Then the ratings started taking off, to 5, 6, and 7. After eight weeks of the 20 milligram dose, they settled between 7 and 8—an impressive increase. Accompanying these high scores were many positive comments in Clara's diary. Particularly noteworthy to me was her newfound tranquility, which had replaced the sadness and agitation she had been feeling when she first came to see me.

But what did these dramatic changes mean? Were they really reflections of a therapeutic change in the chemistry of Clara's brain? Or were they, instead, manifestations of the placebo effect? After all, there were many reasons why Clara might respond positively to a placebo. Among them was her desire to feel better, her desire to prove that her faith in the drug had been justified, and even her desire to please me by showing what a good doctor I was.

The main reason I felt that Clara was not merely responding to Prozac as a placebo was that it took a long time for her to feel better.

It is well known that antidepressants usually don't have much effect until they have been taken for several weeks, whereas placebo effects tend to come on more quickly,[†] often immediately. The long delay before Clara's mood brightened is typical of a true drug effect instead of a placebo effect. Yet, pleased though I was with Clara's improvement, I still was not convinced that I knew what was going on.

Then came a surprise—a surprise so dramatic that it greatly increased my conviction that the drug was really helping Clara. The surprise was a confession by Clara about a secret idea that she had been ruminating about for many years and which brought her intense discomfort. Despite the pivotal significance of this secret idea, Clara had found it so mortifying that she hadn't been able to tell me about it in the dozen or so times I had seen her. Nor had she ever mentioned it to her therapist or to anyone else. Now she was willing to unburden herself. Because, after four months of a full dose of Prozac, this secret idea had undergone a remarkable change: it was almost gone!

The mortifying secret, I was astonished to learn, concerned her nose. One day when Clara was almost fourteen she had somehow come to the conclusion that her long and pointy nose was deformed. Adding to her distress was her feeling that everyone who looked at her was repelled by its ugliness. Since then, Clara had been obsessed by the idea that she had a repugnant nose. Even though her rational self knew that this was not so, thoughts about her nose kept returning, sapping her ability to concentrate on her work. Despite intense efforts to ignore this imagined deformity, Clara spent hours contemplating plastic surgery and kept seeking out mirrors to inspect her appearance. She redid her hair in a style that she devised to divert attention away from her nose.

It is, in fact, a prominent nose, a noticeable and distinctive feature. But when I looked at Clara it was her pale blue eyes, her bright red hair, and her stylish skirts and blouses that caught my eye. Until she made her confession, it had never occurred to me that she had any concerns about her appearance. I found myself wanting to tell her how disappointed I was that she hadn't confided in me from the start.

I was, however, restrained by the realization that this was hardly the first time that a patient with a prominent obsession had kept it secret. And I was consoled by the knowledge that I had gained a better understanding of the nature of Clara's emotional distress. She wasn't simply suffering from depression. Her most distinctive and distressing symptom was obsessive thoughts about her nose. Her main diagnosis wasn't dysthymic disorder.[†] It was something else—either obsessive-compulsive disorder (uncontrollable recurrent thoughts and repetitive behaviors) or a variant called body dysmorphic disorder[†] (a preoccupation with an imagined defect in appearance). Whichever of these terms best characterized Clara's pattern of behavior seemed merely academic, because the main point was clear: despite the fact that I had, until then, failed to discover what was really bothering Clara, I had had the good luck to prescribe a medication that produced dramatic relief. Although I had no idea why Clara was obsessed with her nose and no convincing explanation for Prozac's magical effect, we could both rejoice in the outcome.

What made Clara's response to the drug particularly interesting to me is that there was already substantial evidence of the reduction of obsessive thoughts by another antidepressant, clomipramine (Anafranil, a relative of amitriptyline), which, like Prozac, prolongs the actions of a neurotransmitter called serotonin. Because many patients didn't like some of the side effects of clomipramine, psychiatrists had begun experimenting with off-label Prozac as a remedy for obsessive thoughts (such as Clara's recurrent ruminations about her nose) and compulsive actions (like her repetitive trips to the mirror). Although this had not yet become an established treatment for obsessive-compulsive disorder or body dysmorphic disorder, there were already some reports of its efficacy. This was subsequently confirmed by a series of double-blind placebo-controlled studies not only with Prozac,[†] but also with related drugs that all augment the actions of brain serotonin. As this evidence accumulated, it increased my confidence that Clara was benefiting from a true pharmacological effect.

Clara—skeptic though she claims to be—is even more convinced than I. She has continued to take Prozac since December

1993 and has been very satisfied with the progress of her life. After working for a year as a teacher, she completed her doctoral thesis and received her Ph.D. She now has a faculty position at a women's college not far from the university where she trained, and is very popular with her students. She has been regularly dating a physicist, and they have been talking about getting married.

Nevertheless, Clara doesn't like some of the effects of her medication. The most troubling to her is the feeling that she is less passionate and argumentative than she used to be and that this has blunted her skills as a professional philosopher. This "dull-witted indifference," as she puts it, was markedly aggravated when we transiently experimented with a larger dose of Prozac, 40 milligrams a day, in an attempt to completely eliminate her lingering concerns about her nose. She has also noticed that, even with her usual dose, her interest in sex is diminished, and she rarely has a sexual climax—both frequent complaints of people who take Prozac and related drugs.

To rid herself of the cognitive and sexual side effects of her daily dose of 20 milligrams, and to see if she could do without it, Clara temporarily stopped taking it three years ago, without telling me. But when her obsession about her nose began to intensify after a few weeks without Prozac, she went back to it with renewed conviction. Since then I have been checking up on her every month or two and have added some cognitive therapy to our otherwise free-ranging conversations. We have also experimented with a supplementary medication, as I will explain later. But Prozac has remained the mainstay of her treatment.

Clara usually accepts her prescriptions with thanks and a smile. But now and then she reminds me that, despite my pharmacological and psychological tinkering, she is completely free neither of symptoms nor of side effects. "Don't get me wrong," she once said, carefully choosing her words. "You know I'm pleased with my treatment. But you can't blame me for hoping for something better than Prozac."

2 :: *The First Blockbuster*

Numerous devious pathways[†] led to chlorpromazine. . . .
The logic of a master plan was completely lacking.

—Seymour Kety (1974)

The existence of drugs that can effectively relieve persistent and disabling forms of mental distress, such as Clara's, can be traced back to a series of surprising discoveries that began around 1950. Until then, psychiatrists had very little interest in drugs. The only ones that they regularly prescribed were sedatives, mainly barbiturates.

Of these the most widely used was phenobarbital (Luminal). Introduced in 1912 by Bayer,[†] the company that had made its name by creating aspirin, it was especially popular with doctors in general practice, who prescribed it in small doses as an all-purpose psychiatric drug—an aspirin for the mind. It was also used extensively by psychiatrists in mental hospitals, who gave large doses to agitated or unruly patients to put them to sleep, sometimes for several days.

But despite the extensive use of phenobarbital, there was keen awareness of its limitations. At best it provides only transient relief of mental distress, which often returns in full force when the drug wears off. Furthermore, repeated use gradually produces adaptive changes in the brain, so that progressively higher doses are needed to get the beneficial effect. To make matters worse, some people who are treated with phenobarbital become addicted to it and begin to use it compulsively. Little wonder, then, that psychiatrists were not enthusiastic about phenobarbital or the other sedative drugs of that period, all of which had similar shortcomings.

Their attitude changed in the 1950s with the discovery of a completely different type of medication. Unlike phenobarbital, this new drug improves the mental functioning of certain patients, rather than simply sedating them. Unlike phenobarbital, the gradual changes in the brain that follow from its repeated use makes it progressively more effective rather than less effective. Unlike phenobarbital, it is not addictive. As psychiatrists became convinced of the great value of this new drug, and the other drugs that soon followed, they started dusting off their prescription pads.

::

The discovery of this remarkable medication[†] depended on a series of lucky breaks. They were set in motion in the late 1940s by Henri Laborit, a surgeon in the French navy, when he became interested in the sleep-inducing properties of an antihistamine called promethazine (Phenergan). Although sedation is not a desirable property of a remedy for colds or allergies—which accounts for the current popularity of nonsedating antihistamines such as Claritin—Laborit decided to turn it to his advantage by utilizing promethazine as an aid in anesthesia. In 1949 he reported this off-label use of promethazine in a Belgian surgical journal.

Aware of Laborit's research, Pierre Koetschet, assistant scientific director at Rhône-Poulenc, the French pharmaceutical company responsible for promethazine, initiated a hunt for derivatives with stronger sedative effects. On October 3, 1950, he circulated a memo[†] in which he recommended "chemical work . . . that will provide substances with maximal activity . . . [in] prolonging the action of general anesthetics." Koetschet's goal was to find a better promethazine that would be widely used by surgeons and anesthesiologists.

Based on Koetschet's suggestion, Paul Charpentier, the chemist who had created promethazine a decade earlier from a smelly tar derived from coal, began tinkering with its structure. In December 1950 he made a novel compound, 4650RP, which he submitted for testing in rats. Having had no reason to believe that there would be anything special about 4650RP, Charpentier was excited to learn that the animal tests showed that it is indeed more sedating than promethazine—just what Koetschet had hoped for. Furthermore, the drug seemed sufficiently safe to justify its use in humans. In April 1951 Rhône-Poulenc made it available to doctors for testing under its new name, chlorpromazine.

Rhône-Poulenc's luck did not end with the discovery of the sedating properties of chlorpromazine. The company also had the good fortune to give test samples to psychiatrists. In retrospect this seems like an obvious thing to have done. But in 1951 most psychiatrists were mainly concerned with the psychological aspects of mental illness and, as I have already indicated, turned to medications only as a last resort. It was Laborit's enthusiasm about the many potential uses of the calming effects of chlorpromazine that encouraged a few psychiatrists to try it. It was they who made the astonishing finding that chlorpromazine is much more than a strongly sedating antihistamine.

Among the psychiatric pioneers was Pierre Deniker,[†] who first learned about the drug from his brother-in-law, a friend of Laborit's. Working with his colleague, Jean Delay, Deniker gave chlorpromazine to severely disturbed patients who were long-term residents of St. Anne's Psychiatric Hospital in Paris. Deniker and Delay were amazed to find that the drug has a calming effect on hyperactive patients whose grandiosity, poor judgment, and elevated mood are characteristic features of bipolar disorder (manic-depressive illness). But even more startling were the dramatic changes produced by chlorpromazine in the thoughts and feelings of patients with the most disabling form of mental illness.

Called schizophrenia, it is the mental illness that people find most difficult to understand, because its symptoms seem so bizarre. Other symptoms of mental illness, such as markedly elevated or depressed mood, are akin to feelings that we have all experienced. Even obsessions, such as Clara's, are not that different from the unreasonable concerns about personal appearance that are so common among teenagers. But it is very hard to find anything in common with someone who is convinced that the FBI has planted worms in his brain (a delusion, or false belief) or who keeps hearing criticisms from a real-sounding but imaginary voice (a hallucination, or false perception). Yet it is just such delusions and hallucinations that are the defining features of a mental state called psychosis, which is the hallmark of schizophrenia.

Although schizophrenia seems so strange and foreign, it is actually all around us, affecting about one in a hundred men and women, generally in their youth. In addition to hallucinations and delusions, which are called positive symptoms (positive in the sense of active, not in the sense of favorable), patients with schizophrenia also frequently have negative symptoms such as apathy and withdrawal. In the days before the discovery of chlorpromazine, those afflicted often became increasingly isolated to escape the taunting voices and the powerful evil forces that they believed to be conspiring against them—a series of reactions that were dramatized in the Academy Award-winning motion picture *A Beautiful Mind*. As their social interactions diminished, they became dependent on the supervised environment that only an asylum could provide.

All this changed with the introduction of chlorpromazine. In addition to its immediate calming effect, the drug gradually weakens the intensity of delusions and hallucinations. After months of treatment these symptoms of psychosis may disappear, and patients become less withdrawn. In this state of recovery they are more open to counseling on ways to cope with their illness. In many cases, this makes possible a return to their families and the resumption of productive lives—as long as they continue taking their medication.

News of these wondrous early findings of the French psychiatrists quickly spread to America, where Smith, Kline, & French, an ambitious Philadelphia pharmaceutical company, obtained the license to market chlorpromazine. By 1955 it was being administered to hundreds of thousands of people with schizophrenia[†] and other mental disorders, under the European trade name Largactil (for "large number of actions") and the American trade name Thorazine. Just four years after its creation by Charpentier, chlorpromazine had become the first blockbuster psychiatric drug.

When psychiatric hospitals started buying chlorpromazine by the truckload in 1955, no one had any idea how it worked. It was clear from the start that interference with the action of histamine is not responsible for the antipsychotic effect, because promethazine, the powerful antihistamine from which chlorpromazine was derived, does not alleviate the symptoms of schizophrenia. Therefore chlorpromazine's antipsychotic effect must be due to some other property of this drug. Identifying that property proved to be as significant as the discovery of chlorpromazine's usefulness for schizophrenia, because it introduced a way of thinking about psychiatric drugs—an important step on the road to other medications such as Prozac.

The first clue came from an additional, and very troublesome, feature of chlorpromazine: when patients take large doses, many develop abnormal movements of the hands and mouth as well as muscular rigidity of the body and face. These symptoms resemble those of a neurological disorder first described in 1817 by James Parkinson, a British physician, and now called Parkinson's disease. This striking side effect of the drug raised the possibility that studies of the origins of Parkinson's disease might also reveal chlorpromazine's secret.

The importance of the relationship between chlorpromazine's effects on psychosis and on body movements was underscored by studies with another drug, reserpine. Derived from snakeroot, a plant that folk doctors in India had been dispensing for centuries as a remedy for agitation, reserpine was widely used in the 1950s as a treatment for high blood pressure. In 1954, two years after the dramatic findings with chlorpromazine, Nathan Kline, a New York psychiatrist, reported that reserpine could also be used to treat schizophrenia. Although reserpine did not prove to be as effective as chlorpromazine, it shared its worst feature: it too produces symptoms like those of Parkinson's disease.

We now know that the symptoms produced by chlorpromazine and reserpine, like the symptoms of Parkinson's disease itself, are due to a decrease in brain signaling by a chemical called dopamine. Much of the credit for solving this mystery in the 1950s and 1960s goes to Arvid Carlsson,[†] a Swedish pharmacologist who had already discovered that dopamine is a neurotransmitter, a chemical that is used to transmit signals between nerve cells. In the course of this research Carlsson found that reserpine and chlorpromazine interfere with the function of dopamine in two different ways: reserpine by reducing the amount of dopamine stored in the brain, and chlorpromazine by blocking the actions of dopamine at its natural targets, called receptors. This body of knowledge has been so influential that it brought Carlsson a share of a 2000 Nobel prize. From it came many important medications, including L-dopa, which the body converts to dopamine, as a treatment for Parkinson's disease.

Even more important was the broader impact of this work. By showing that chlorpromazine works by interfering with the action of a neurotransmitter, it brought other neurotransmitters—such as norepinephrine, serotonin, acetylcholine, glutamate, and gamma-aminobutyric acid (GABA)—to center stage. From then on research on psychiatric drugs has been intimately linked with research on neurotransmitters.

::

Neurotransmitters are the chemicals that nerve cells use to communicate with each other[†]—the substances that are continuously shaping our mental processes. Every thought and feeling depends on the release of neurotransmitters by billions of nerve cells, and the reception of these chemical signals by billions of other nerve cells. The structures where this communication occurs are points of contact between nerve cells, called synapses. To understand the actions of psychiatric drugs, it is necessary to understand the workings of synapses.

Each synapse is made up of parts of two different nerve cells that are separated by a small space called the synaptic cleft. The specialized part of one nerve cell, called the nerve terminal, manufactures a neurotransmitter, such as dopamine. When this nerve cell is activated by signals from other nerve cells, it squirts some of its neurotransmitter into the synaptic cleft. The neurotransmitter floats over to the surface of the other nerve cell, where it binds to its receptors, which are specialized proteins with grooves that recognize the neurotransmitter. For example, receptors with grooves that recognize dopamine are called dopamine receptors.

When a neurotransmitter is released at a synapse and binds to a receptor on the surface of a nerve cell, the bound receptor sends a signal into the interior of the nerve cell, setting in motion a complex series of processes. Certain bound receptors send activating signals (also called excitatory signals), whereas others send calming signals (also called inhibitory signals). The combined effects of tiny activating and calming signals in the trillions of synapses in the brain are integrated into the emotions, perceptions, and thoughts that we experience at a given moment.

The nature of the contribution of an individual synapse is determined by the function of the circuits of nerve cells in which it participates. For example, certain synapses that use dopamine reside in several large clusters of nerve cells in the core of the brain, called the basal ganglia, which play a role in the control of movement. It is the degeneration of the nerve terminals that release dopamine at these synapses that is responsible for the abnormal movements of people with Parkinson's disease, and it is chlorpromazine's blockade of the grooves of dopamine receptors in synapses of the basal ganglia that produces similar symptoms. Other synapses that use dopamine are distributed in regions of the brain such as the frontal cortex and the limbic system, which control thoughts and emotions. These synapses also contain dopamine receptors whose grooves are blocked by chlorpromazine. It is the blockade of these synapses by chlorpromazine that seemed most likely to be responsible for its therapeutic effects on psychosis.

But knowing that chlorpromazine blocks the grooves of dopamine receptors doesn't prove that it is this same action that relieves the abnormal thinking of people with schizophrenia. Just as chlorpromazine's blockade of receptors for histamine was shown to be irrelevant—because promethazine, a related antihistamine, does not share its antipsychotic effect—so too might chlorpromazine's blockade of receptors for dopamine be irrelevant.

The way this matter was settled was by studying the new antipsychotic drugs that had been developed by drug companies in the wake of chlorpromazine's discovery. Of these new drugs, several that became very popular in the 1960s are far more potent than chlorpromazine—which means that smaller amounts are sufficient to relieve delusions and hallucinations. One of them, fluphenazine (Prolixin), was made from chlorpromazine by modifying its chemical structure. Another, haloperidol (Haldol), a different type of chemical that was discovered because it mimics the effects of chlorpromazine in mice, was introduced in 1959 by Janssen, a Belgian drug company, as an alternative treatment for schizophrenia. If the antipsychotic effect of chlorpromazine is truly related to its anti-dopamine effects, these more potent antipsychotics should also be more potent blockers of the binding of dopamine to its receptors.

This is just what was found by groups of scientists led by Solomon Snyder at Johns Hopkins University and by Philip Seeman at the University of Toronto, who devised methods for measuring binding of dopamine to receptors[†] in tiny bits of brain tissue. Compared with chlorpromazine, much smaller amounts of haloperidol or fluphenazine were needed to block a given amount of binding of dopamine to its receptors. The correlation of antipsychotic potency with the degree of blockade of dopamine binding supported the proposal that the blockade is responsible for the antipsychotic effects.

Further support for this idea came from two discoveries about amphetamine, a drug that was introduced in the 1930s (which I will return to later), and which is prone to abuse. First it was observed that prolonged bingeing on large doses of amphetamine frequently results in intense paranoia and other behavioral changes that resemble the positive symptoms of schizophrenia. Later it was found that the main action of amphetamine is to pump dopamine from nerve terminals into synapses. When combined with the evidence that antipsychotic drugs block neurotransmission by dopamine, the data about amphetamine gave rise to a proposal about the origin of schizophrenia, called the dopamine hypothesis of schizophrenia.[†] In this view the symptoms of schizophrenia are the result of excessive neurotransmission by dopamine in the parts of the brain that control thoughts and emotions—the same sort of idea about a chemical imbalance that persuaded Clara to try Prozac.

It is not, of course, that simple, because chlorpromazine and other antipsychotic drugs block dopamine receptors within minutes after they are taken, whereas their reduction of delusions and hallucinations develops over many weeks.[†] This implies that by continuously blocking dopamine receptors, the drugs also set in motion a gradual series of changes in the brain that somehow counteract the psychotic process, whether or not it was caused by too much dopamine. In fact, there is no direct evidence that schizophrenia is caused by excessive dopamine neurotransmission. Nevertheless, the dopamine hypothesis dominated thinking about this psychotic state for many years.

::

What helped modify this simple view was another lucky break: the discovery of the special therapeutic properties of an antipsychotic drug called clozapine (Clozaril). Introduced in Europe in the 1960s, clozapine seemed at first like just another me-too medication that pharmaceutical companies kept trying to promote as replacements for chlorpromazine. Although it was clear from the start that clozapine is an effective treatment for schizophrenia, enthusiasm waned when a few people who took it developed life-threatening infections due to the loss of a type of white blood cells called granulocytes. As a result, its manufacturer, Sandoz, began taking clozapine off the shelves, and many of the patients who were taking it were switched to other antipsychotic drugs.

But to everyone's surprise, some of those who had been helped by clozapine relapsed when it was replaced with other medications, suggesting that it has a unique therapeutic property. This raised the possibility that clozapine could also bring relief to some of the other patients who weren't helped by any of the standard antipsychotic drugs. When this was shown to be the case in 1988,[†] clozapine was again made available, but this time only to a select group of patients who agreed to have weekly blood tests to check their granulocytes.

Should the number of granulocytes fall to a dangerous level, the drug could be discontinued in time to prevent a potentially fatal infection.

Enthusiasm for clozapine grew when it became clear that it may even have advantages for those patients who had a favorable response to chlorpromazine, fluphenazine, or haloperidol. Whereas all these early drugs—now called typical antipsychotics—can control delusions and hallucinations, many patients who are treated with them still have negative symptoms such as apathy and social withdrawal. For some of these patients, switching to clozapine increases sociability and zest for living.

In addition to these psychological benefits, clozapine has another advantage: patients treated with this drug rarely develop Parkinson's-disease-like symptoms. This finding added a new wrinkle to the correlation between therapeutic efficacy for schizophrenia and interference with neurotransmission by dopamine. Although clozapine does indeed block certain actions of dopamine, the effect is relatively more prominent in the regions of the brain that control thoughts and emotions (the frontal cortex and limbic system) than in the regions that control movements (the basal ganglia).

The selective effect of clozapine on dopamine functions in different parts of the brain may explain why it lacks yet another damaging side effect of the typical antipsychotics. Called tardive (late-occurring) dyskinesia (abnormal involuntary movements), it becomes quite common when these old-line drugs are taken for many years. Tardive dyskinesia, which includes involuntary grimacing and tongue-thrusting movements, is particularly embarrassing to patients, because observers may take it as a sign of insanity. Believed to be an idiosyncratic adaptation of the brain to a sustained blockade of dopamine neurotransmission, tardive dyskinesia is the most hated side effect of psychiatric drugs, because it can persist indefinitely even after stopping the medication. In the days when clozapine was the only alternative to the typical antipsychotics, many patients were willing to tolerate the inconvenience of weekly blood tests simply to escape the danger of developing tardive dyskinesia.

The advantages of clozapine—absence of Parkinson's-disease-like symptoms, absence of tardive dyskinesia, and a reduced likelihood of apathy and social withdrawal—became defining features of a new category of drugs for schizophrenia, called atypical antipsychotics. Once this category was recognized, many drug companies became interested in finding a substitute for clozapine that does not harm white blood cells. Just as the accidental discovery of the antipsychotic properties of chlorpromazine led to a race to find more potent variations, so too did the accidental discovery of the special benefits of clozapine lead to a race to find others like it. The search for new drugs for schizophrenia was on.

::

The initiation of this search for atypical antipsychotics in the late 1980s coincided with the discovery that the brain may have a variety of receptors for each neurotransmitter. For dopamine there are five receptors, called D_1 through D_5, that are distributed in particular parts of the brain and produce different effects on the cells that contain them when they are activated by dopamine. Once these and other neurotransmitter receptors were identified, techniques were developed to examine the interaction of existing psychiatric drugs with each of them. The relative binding of these drugs to different receptors became a basis for designing new ones.

It quickly became clear that all antipsychotic drugs bind to and change the properties of many neurotransmitter receptors. For this reason they are sometimes called "dirty drugs" to contrast them with "clean drugs" that are more selective. The dirtiness of antipsychotic drugs is not surprising because many neurotransmitter receptors have similar protein structures and would therefore be expected to share an affinity for certain chemicals. Nevertheless, each receptor has relatively strong preferences for some drugs and lesser preferences for others. For example, clozapine binds relatively well to the D_4 receptor, whereas haloperidol and other typical antipsychotic drugs bind better to the D_2 receptor. Clozapine also binds relatively well to certain serotonin receptors and to some other receptors.

Although a detailed study of clozapine's binding to receptors did not reveal all the reasons for its unusual properties, it contributed to the development of new atypical antipsychotics. By examining the binding of a wide range of potential drugs to a series of isolated receptors, pharmacologists found some that shared clozapine's pattern of binding to serotonin receptors and dopamine receptors. This line of research led to the introduction of several effective antipsychotic drugs that have many of the favorable features of clozapine[†] but don't harm white blood cells. These drugs are now replacing the older antipsychotic medications. They are even being substituted for clozapine in the treatment of some patients who have benefited greatly from that drug but are tired of the repeated blood tests.

The first of the new atypical antipsychotics to be marketed was risperidone (Risperdal). Although it is not completely free of Parkinson's-disease-like effects, risperidone has been a huge success. It was soon followed by olanzapine (Zyprexa), which more closely resembles clozapine and which shares two of its major drawbacks: it makes people fat and may cause diabetes. Despite these limitations, Zyprexa has become another blockbuster.

Eager to share in the growing market for atypical antipsychotics, other drug companies introduced alternatives with different patterns of side effects. Of these, quetiapine (Seroquel) is being promoted as an antipsychotic that is effective at doses that don't produce any Parkinson's-disease-like symptoms. The most recent entry in this new battle of the side effects is ziprasidone (Geodon), which has not produced obesity in clinical trials. But, as its competitors are eager to point out, Geodon can produce potentially dangerous changes in heart rhythm.

Presently a new group of antipsychotic drugs is being developed that some refer to as atypical atypicals. What distinguishes these drugs from their predecessors is that they share certain properties of both chlorpromazine and dopamine. In some circumstances they are blockers (or antagonists) of dopamine receptors—like chlorpromazine. In other circumstances they are activators (or agonists) of dopamine receptors—like dopamine itself. Also called partial dopamine agonists[†] or dopamine system stabilizers,[†] they are effective treatments for schizophrenia.

The first member of this group to be introduced is aripiprazole. It was developed in Japan[†] by Otsuka Pharmaceuticals and is being marketed in the United States by Bristol-Myers Squibb under the trade name Abilify. So far aripiprazole appears to have fewer distressing side effects than its more established competitors. But it is too early to tell how useful it will be.

::

Despite their many valuable features, the therapeutic effects of all the new antipsychotic drugs are mainly based on their interaction with dopamine receptors, especially the D_2 receptor. But receptors for other neurotransmitters have also been implicated in psychosis, raising the possibility that they too could become targets for novel antipsychotic drugs. The most notable example is the NMDA receptor,[†] one of the receptors for a neurotransmitter called glutamate, which is the brain's most widely used excitatory (activating) neurotransmitter.

The NMDA receptor is a complex protein that can bind a variety of chemicals. Among them is a drug called phencyclidine, which blocks this receptor and diminishes neurotransmission by glutamate. Developed for use as an anesthetic in the 1950s, phencyclidine was withdrawn in 1965 because of its many undesirable properties and is now only used illicitly (known as PCP or angel dust). Among phencyclidine's effects are hallucinations and a sense of detachment[†] that, together, resemble the whole range of positive and negative symptoms of schizophrenia. This raised the possibility that diminished glutamate neurotransmission may cause some of the symptoms of schizophrenia, and stimulated a search for antipsychotic drugs that work by augmenting the action of glutamate.

But stimulating glutamate receptors is potentially risky, because a substantial increase in glutamate neurotransmission is known to cause nerve cell damage. To gently increase glutamate neurotransmission, attention has focused on a specialized part of the NMDA receptor that binds glycine or D-serine.[†] These brain chemicals can each augment the response of the NMDA receptor to glutamate without activating other glutamate receptors.

To evaluate the therapeutic effects of glycine and D-serine, each has been given to patients with schizophrenia. When given alone, neither of them produced any benefit. But when given to patients who were already taking antipsychotic drugs such as chlorpromazine or haloperidol, D-serine produced some additional improvement.[†] Although the results were not very impressive, they have encouraged pharmaceutical companies to search for better ways to selectively augment glutamate neurotransmission at NMDA receptors.

::

Their investment in this innovative approach is, however, quite limited. Pharmaceutical companies have learned through long experience that they get a much better financial return by creating and testing modifications of established medications, as in the case of the atypical antipsychotics. They have also learned that, in addition to its stodgy practicality, this strategy of testing such me-too drugs can sometimes lead to wonderful surprises. And this, as I will now explain, is just what happened in the very first search for a replacement for chlorpromazine.

3 :: The Road to Prozac

Not infrequently the cure [produced by imipramine] is complete, sufferers [of depression] and their families confirming the fact that they had not been so well for a long time.
—Roland Kuhn (1958)

When Rhône-Poulenc made a splash with chlorpromazine as a treatment for schizophrenia, others decided to try to make competitive products. Among them was Geigy, a Swiss pharmaceutical company, which rummaged through its storehouse for similar chemicals that had originally been prepared in a search for antihistamines. Hoping to find an antipsychotic drug of its own, Geigy chose G22355, the chemical that most closely resembled chlorpromazine, and distributed it for testing. Among the psychiatrists who received a supply of G22355 was Roland Kuhn at the state psychiatric hospital in Münsterlingen, Switzerland.

The outcome was disappointing. G22355 was clearly not an antipsychotic. Instead of calming patients with schizophrenia and stopping their delusions and hallucinations, G22355 sometimes increased agitation and made the psychotic symptoms worse. Geigy was ready to throw in the towel.

But as with the early tests with chlorpromazine, G22355 also had an unexpected property—in this case elevation in the mood of some of the gravely ill patients who received it. This led Kuhn to give the drug to a woman who had been hospitalized because of severe depression. The result was astounding. With just one week of treatment there was obvious improvement.

Good results were also observed in two other depressed patients. As Kuhn later recalled,[†] "After treating our first three cases, it was already clear to us that the substance G22355, later known as imipramine, had an antidepressive action. On February 4, 1956, we sent Geigy a long report in which we pointed out the effect this substance had in depressions. Specifically, we said that if this finding were confirmed it would be of the utmost practical importance."

::

Confirmation soon came. By 1958 Kuhn had given imipramine to several hundred patients with major depression—a form of depression that is much more severe than Clara's—and had many excellent results. Major depression is remarkably common, affecting at least one in twenty people at some time in their lives, often repeatedly. Because it is so prevalent and can go on for many months at a time, the World Health Organization ranks major depression second only to heart disease as a source of suffering and disability.

It often starts gradually and then, for no apparent reason, gains in intensity. Writing about his depression in the *New Yorker* in January 1998, Andrew Solomon, who went on to publish *A Noonday Demon,* an award-winning book on this subject, describes his progressive slide:

> In June 1994, I began to be constantly bored. My first novel had been recently published in England, and yet its favorable reception did little for me. I read the reviews indifferently and felt tired all the time. In July, back home in downtown New York, I found myself burdened by phone calls, social events, conversation. The subway proved intolerable. In August, I started to feel numb. I didn't care about work, family, or friends. My writing slowed, then stopped. My usually headstrong libido evaporated. . . .
>
> Sleeping pills got me through the night, but morning began to seem increasingly difficult. From then on, the slip-

page was steady. I worked even less well, cancelled more plans. I began eating irregularly, rarely feeling hungry. A psychoanalyst I was seeing told me, as I sank lower, that avoiding medication was very courageous.

At about this time night terrors began. My book was coming out in the United States, and a friend threw a party on October 11th. . . . The event lives in my mind in ghostly outlines and washed-out colors. When I got home, terror seized me. I lay in bed, not sleeping and hugging my pillow for comfort. Two weeks later—the day before my thirty-first birthday—I left the house once, to buy groceries; petrified for no reason, I suddenly lost bowel control and soiled myself. I ran home, shaking, and went to bed, but I did not sleep, and could not get up the following day. I wanted to call people to cancel birthday plans, but I couldn't. . . . At about three that afternoon, I managed to get up and go to the bathroom. I returned to bed shivering. Fortunately my father, who lived uptown, called about then. "Cancel tonight," I said, struggling with the strange words. "What's wrong?" he kept asking, but I didn't know.

Although Kuhn also hadn't known what was wrong with patients like Andrew Solomon, he was very pleased to find that quite a few of them were helped by imipramine. In his classic 1958 paper, he published the following description of his results:[†]

The effect is striking in patients with a deep depression. We mean by this a general retardation in thinking and action, associated with fatigue, heaviness, feeling of oppression, and a melancholic or even despairing mood, all of these symptoms being aggravated in the morning and tending to improve in the afternoon and evening. From the external appearance alone it is possible to tell that the mood improves with imipramine hydrochloride. The patients get up in the morning of their own accord, they speak louder and more rapidly, their facial expression becomes more vivacious. They commence some activity of their own, again seeking contact with other people, they begin to entertain themselves, take part in games, become more cheerful, and are once again able to laugh. . . . Instead of being concerned about imagined or real guilt in their past, they become occupied with plans concerning their future. . . . Suicidal tendencies also diminish, become more controllable or disappear altogether. . . . Where the depression was accompanied by insomnia sleep occurs again . . . and the sleep is felt to be normal and refreshing, not fatiguing and forced as that so often produced by sleeping remedies.

Kuhn's report was soon confirmed by other psychiatrists. Accustomed to feeling powerless in the face of the incomprehensible despair of their patients with major depression, they were thrilled that a useful remedy had been found at last. By 1960 imipramine was flying off the shelves under the trade name of Tofranil.

::

While Kuhn was conducting these groundbreaking studies with imipramine, observations were being made about the antidepressant properties of a completely different type of drug called iproniazid. Introduced in 1951 for the treatment of tuberculosis by Hoffman–La Roche, another Swiss pharmaceutical company, doctors who prescribed iproniazid were surprised to find that this antibiotic affected the behavior of their patients. Some became irritable and confused when they took large doses, and had to be switched to alternative medications for tuberculosis. But others became pleasantly activated and reported that their mood was improved.

Learning of these favorable effects on mood, Nathan Kline, who had already shown that one established drug—the blood pressure medicine reserpine—could be used for the treatment of schizophrenia, became excited about the possibility of trying another established drug—the tuberculosis medicine iproniazid—to treat apathetic or depressed psychiatric patients. Because pharmacies were stocking iproniazid for the treatment of tuberculosis, it was easy for Kline and his collaborators to prescribe it off-label. Based on limited evidence, Kline concluded that the drug was useful for depression. In 1957, while Geigy was gearing up to release imipramine, iproniazid was already being used widely[†] as an antidepressant.

The demonstration that iproniazid can relieve depression was particularly interesting because scientists had found that this antibiotic has an additional and unrelated property: it inactivates an enzyme in the brain. Called monoamine oxidase (abbreviated MAO), this enzyme destroys the norepinephrine, serotonin, and dopamine that are squirted out of nerve terminals in the process of neurotransmission. This destruction of neurotransmitters is one way of terminating signals in a brain circuit, thereby clearing it for another communication.

Inactivation of MAO by iproniazid prolongs signals by stopping destruction of the neurotransmitters. It also increases the amounts of norepinephrine, serotonin, and dopamine that are stored in nerve terminals, so more is at hand for neurotransmission. The net result is a complex readjustment in the activity of many circuits in the brain.

The change produced by iproniazid—increases in norepinephrine, serotonin, and dopamine—is the opposite of the one produced by reserpine. That drug blocks the storage not only of dopamine, as described earlier, but also of norepinephrine and serotonin, leading to a depletion of all three of these neurotransmitters. Furthermore, these opposite effects of iproniazid and reserpine on brain chemistry are correlated with observations of their opposite psychological effects: iproniazid elevates mood, whereas reserpine depresses it. When taken at face value these findings suggest that low levels of norepinephrine, serotonin, and dopamine produce depression, whereas high levels of these neurotransmitters relieve depression.

::

One thing that did not seem to fit with this view was that imipramine, a potent antidepressant, doesn't change the breakdown of neurotransmitters by enzymes such as MAO. How, then, does imipramine relieve depression? Can imipramine's effect be reconciled with the claim that depression is caused by a chemical imbalance?

The solution to this puzzle came from the National Institutes of Health (NIH) in Bethesda, Maryland, which had already contributed many of the early ideas about the interaction of drugs with brain chemicals. The godfather of this stimulating scientific milieu was Bernard Brodie,[†] a pharmacologist who had made his reputation during World War II by creating new drugs for malaria, then turned his attention to the effects of drugs on the brain. It was in Brodie's laboratory that the depletion of brain serotonin by reserpine was discovered, and that speculations about the psychological effects of this neurotransmitter were most actively discussed. It was in Brodie's laboratory that Arvid Carlsson learned the methodology that made it possible for him to discover dopamine and its relationship to antipsychotic drugs, which eventually won him a Nobel prize. And it was in Brodie's laboratory that another future Nobel laureate, Julius Axelrod, was set on a path that led him to figure out the secret of imipramine.

Having started in Brodie's laboratory as a technician, Axelrod was already a highly accomplished scientist when he belatedly got his Ph.D. While establishing a research program of his own, Axelrod focused his attention on norepinephrine and discovered an enzyme, called catechol-o-methyltransferase (abbreviated COMT), that inactivates this neurotransmitter in a different way than MAO, also helping to terminate its action. Then, in the course of further work with norepinephrine, Axelrod made an even more remarkable discovery:[†] that the action of the norepinephrine that had been squirted out into a synapse can be terminated by simply sucking the neurotransmitter back into the nerve terminals from which it originated. The beauty of this alternative way of terminating the chemical signal—called reuptake—is that the norepinephrine is not degraded in the process, so it can be used again.

Once Axelrod had established this novel mechanism for terminating the action of norepinephrine, he began looking for drugs that could block this process, a quest whose successful result I happened to be among the first to learn about. Having just started an apprenticeship in research at NIH in 1960, I frequently visited Axelrod and his trainees to get advice for a study of the effects of norepinephrine and serotonin on the pituitary gland. In one of my visits to Axelrod's lab, which was just a few floors below mine in NIH's massive Building 10, I got the word that Julie was on to something very important. This didn't surprise me at the time because it seemed to me that Julie was always on to something very important. Only when I was told that he had discovered that imipramine blocked reuptake of norepinephrine did I understand why the people in his lab were so excited. Axelrod's 1961 paper describing this discovery has played a critical role in our understanding of neurotransmission and of the actions of a variety of psychiatric drugs, and set the stage for his 1970 Nobel prize.

We now know that the reuptake process for terminating the action of norepinephrine that Axelrod discovered is controlled by a protein called the norepinephrine transporter. Deployed on the surface of nerve terminals that release this neurotransmitter, the protein is poised to grab norepinephrine and transport it back into the terminal. We now also know that imipramine interferes with this transporter, inhibiting reuptake and allowing the neurotransmitter to keep working for a longer period of time. So the net result of taking imipramine and blocking the reuptake of norepinephrine resembles that of taking iproniazid and blocking the degradation of norepinephrine by MAO. In both cases there is a prolongation of the action of norepinephrine—an alteration in the chemical balance of the brain.

::

Axelrod's demonstration that imipramine, like iproniazid, may augment neurotransmission by norepinephrine had two important implications. First, it raised the obvious possibility that the therapeutic effect of both antidepressants is due to a drug-induced increase in neurotransmission in brain circuits controlled by norepinephrine. Second, it supported the growing idea of chemical imbalance by inviting the speculation that the symptoms of depression might actually be due to a deficiency of norepinephrine in the brain.

Together these two inferences became the norepinephrine hypothesis of depression[†]—the first detailed elaboration of the idea that mental disorders are due to a chemical imbalance in the brain, and the precursor of ideas about the role of dopamine in schizophrenia that I mentioned earlier. Put simply, the norepinephrine hypothesis of depression states that depression is due to a deficiency of norepinephrine, and that imipramine and iproniazid relieve depression by correcting the deficiency.

This view of depression was popularized by Joseph Schildkraut, William Bunney, and John Davis, three young psychiatrists who had come to NIH to do research on mental disorders in a milieu enriched by scientists such as Brodie and Axelrod. In 1965 they published papers in the *Archives of General Psychiatry* and the *American Journal of Psychiatry,* the two most influential journals in the field, in which they reviewed the evidence for the norepinephrine hypothesis of depression. These papers and a 1967 follow-up in *Science* by Schildkraut and his mentor, Seymour Kety, stimulated enormous interest in this idea.

But even before these papers were published, there was already reason to believe that norepinephrine should be prepared to share the limelight with a fellow neurotransmitter—serotonin. After all, serotonin is also affected by iproniazid and reserpine. Furthermore, beginning in 1963, Julius Axelrod and others showed that imipramine blocks reuptake of serotonin[†] in the same way that it blocks reuptake of norepinephrine. So all the arguments for a norepinephrine hypothesis of depression could be reformulated to also support a serotonin hypothesis of depression.

But further studies of imipramine reaffirmed the importance of norepinephrine. What temporarily tipped the balance in its favor was the discovery that it was not just imipramine that was clearing up depression but also a modified chemical called desipramine. This modification is made in the liver, which converts imipramine to desipramine as soon as the contents of the pill are absorbed into the bloodstream, so that it is mainly desipramine, rather than imipramine, that reaches the brain. Yet unlike imipramine, which blocks reuptake of both norepinephrine and serotonin, desipramine only blocks reuptake of norepinephrine.

Additional support for norepinephrine came from similar studies with amitriptyline—Clara's first drug—which is chemically related to imipramine, and which Merck introduced in 1961 as Elavil, an antidepressant. Like imipramine, amitriptyline blocks reuptake of both norepinephrine and serotonin. Like imipramine, amitriptyline is rapidly modified by the liver, in this case to nortriptyline. And like desipramine, nortriptyline blocks reuptake of norepinephrine but not serotonin. Furthermore, neither of these modified forms has much effect on the reuptake of dopamine, underscoring their specificity for the norepinephrine transporter.

Once these modifications were discovered, drug companies began marketing desipramine and nortriptyline as antidepressants. Their finding that both are as effective as the parent compounds removed any lingering doubts that imipramine and amitriptyline can do their job simply by blocking reuptake of norepinephrine.

::

As experience accumulated with imipramine, amitriptyline, and their derivatives—collectively called tricyclic antidepressants, or TCAs—it became clear that many patients stopped taking them because they cause constipation, dry mouth, heart irregularities, and other unpleasant side effects. Patients also disliked the side effects of the MAO inhibitors, the other major class of antidepressants, such as tranylcypromine (Parnate), that were developed as improved versions of iproniazid. The most notable is the development of very high blood pressure after eating a variety of foods, especially aged cheeses, that are rich in tyramine, a monoamine that is normally detoxified by MAO. Because patients balked at the dietary restrictions that accompanied prescriptions for MAO inhibitors, and hated the side effects of the TCAs, a search began for antidepressants that didn't have these drawbacks.

One of the leaders of this search was Arvid Carlsson, the Swedish pharmacologist who had discovered dopamine. Starting with yet another antihistamine, brompheniramine, Carlsson and his colleagues found that, like imipramine, this drug blocked reuptake of norepinephrine and serotonin. The surprise came when they tested a derivative that had been prepared by chemists at Astra, a Swedish pharmaceutical company. Unlike its parent, brompheniramine, the derivative, named zimelidine, blocked reuptake of serotonin but not norepinephrine. Furthermore, zimelidine was not converted by the liver into an inhibitor of norepinephrine reuptake. Yet, despite its lack of effect on norepinephrine, zimelidine—the first selective serotonin reuptake inhibitor (SSRI)—relieved depression.[†]

In 1982 Astra began selling zimelidine in Europe as an antidepressant with the trade name of Zelmid. Licensed by Merck for distribution in the United States, Zelmid might well have been on its way to becoming a breakthrough drug. But then disaster struck: several patients who were treated with Zelmid developed a form of paralysis called Guillain-Barré syndrome. Although this side effect is rare, it can persist after the drug is discontinued, and is potentially fatal. As a result, Zelmid was withdrawn from the market.

Astra's cloud had a silver lining. Its beneficiary was Eli Lilly, an American pharmaceutical company based in Indiana. In the early 1970s it too had started making derivatives of an antihistamine, in this case diphenhydramine—the active ingredient in popular allergy pills such as Benadryl, as well as over-the-counter sleeping pills such as Sominex and TylenolPM. Explicitly interested in finding a new antidepressant, scientists at Lilly tested the effects of the new chemicals on reuptake of serotonin and norepinephrine. On July 24, 1972, David Wong, a Hong Kong–born pharmacologist, found that one of these chemicals, Lilly 110140, is a potent inhibitor of serotonin reuptake[†] and a weak inhibitor of norepinephrine uptake. Named fluoxetine, this selective serotonin reuptake inhibitor would eventually become world-famous—as Prozac.

But fluoxetine's great promise was not immediately apparent. In the decade after Wong's discovery, Lilly was reluctant to invest the vast sum needed to find out if it could be used to treat depression. Learning of Astra's success with zimelidine, Lilly even considered setting fluoxetine aside and vying with Merck for the U.S. rights to

Zelmid. Were it not for zimelidine's rare but potentially fatal side effect, the household word for antidepressant would now be Zelmid instead of Prozac.

::

Getting Prozac approved for sale as an antidepressant was not, however, as easy as Lilly would have liked. In the time since imipramine and amitriptyline were introduced, based largely on the observations of experienced psychiatrists such as Kuhn, the rules had changed. Stimulated by the discovery of the disastrous malformations produced by thalidomide, a sleeping pill that had been marketed in Germany without adequate testing, the U.S. Congress passed a law in 1962 that increased the stringency of the review of new drugs. Called the FDA Amendment of 1962, it instructed the Food and Drug Administration to approve new drugs only if they had been clearly shown to be both safe and effective. Safety would be determined by extensive tests in animals, followed by tests in human volunteers. Effectiveness for the treatment of a particular disorder would be determined not just by the observations of experts but also by formal comparisons with sugar pills in randomized controlled trials.

The need for careful evaluation of the effectiveness of antidepressants was already apparent in the initial study of imipramine. Although Kuhn was convinced that this drug could produce miracles, he readily acknowledged its variable effects in his famous 1958 paper:

> Not infrequently the cure is complete, sufferers and their relatives confirming that they had not been so well for a long time. . . . In many cases however, there is merely some degree of improvement, making the condition more bearable for the patient, and even permitting resumption of work, though at the cost of considerable effort. In other cases there is no effect at all.

But what exactly is "some degree of improvement"? And how much of this improvement is attributable to a placebo effect rather than a specific therapeutic action of the drug? Whereas Kuhn was content to make qualitative estimates of the effects of drugs on the course of depression in individual patients, the FDA Amendment of 1962 led to a demand for quantitative estimates of the effects of both drugs and placebos on the course of depression in groups of patients.

A way to meet the FDA demand had already been provided in 1960 by Max Hamilton, of the University of Leeds, who devised a rating scale for depression[†] that was specifically designed to assess the results of treatment. Called the Hamilton Depression Scale (Ham-D), it consists of seventeen items that are scored by a professional in the course of interviewing a patient. Some of the items, such as depressed mood, feelings of guilt, and risk of suicide, are each scored on a scale of 0 to 4, where 0 means "absent" and 4 is "severe." Other items such as agitation are scored on a scale of 0 to 2, which gives them less weight in the final tally. The highest possible score is 52. Scores above 30 are generally taken to mean that the person is severely depressed. People with scores of 7 or less are considered to be normal.

As with all psychiatric rating scales, the one introduced by Hamilton provides no more than a rough estimate of the severity of the condition that it measures. Nevertheless, because of its timely introduction, the Ham-D has become the standard for evaluating treatments for depression. When alternative scales are used to examine the effects of antidepressants, the results are similar.

Using these scales to evaluate the effects of imipramine on patients with major depression, the results resembled those that Kuhn had reported in 1958. As Kuhn had observed, many weeks of drug treatment are often needed for an individual to get a maximal therapeutic effect, and some see no effect at all. This has been confirmed in studies of severely depressed patients[†] with initial Ham-D scores of about 25 to 30. After two months of imipramine treatment the scores of about half of the patients had declined by at least 50 percent (e.g., from 25 to 12), which corresponds with what Kuhn called "some degree of improvement." But few of them had Ham-D scores below 7, which is generally taken as the cutoff point that indicates full recovery. And many showed little or no improvement, corroborating Kuhn's observation that not all depressed patients are helped by imipramine.

What Kuhn had not anticipated was that some depressed patients who received sugar pills for two months were also feeling a lot better[†] than they had at the start. So even though Kuhn's observations had been correct, only a portion of the improvement he observed was attributable to the effect of imipramine on brain chemistry. As with many popular treatments for nonpsychiatric disorders, a part of the therapeutic effect can be mimicked by a pharmacological or even surgical placebo.[†]

Similar results were found in studies of Prozac and the other SSRIs—sertraline (Zoloft), paroxetine (Paxil), fluvoxamine (Luvox), and citalopram (Celexa)—which came to market in its wake. They were also found with still other antidepressant drugs such as venlafaxine (Effexor), mirtazapine (Remeron), nefazodone (Serzone), and bupropion (Wellbutrin), which have a variety of effects on neurotransmission.[†] All these drugs beat out placebo and are about as effective as imipramine. The other consistent finding is that none of them works for all patients. About one out of three people with major depression don't get much benefit from the first antidepressant drug they are given. Some are helped by another antidepressant,[†] by certain combinations of two of them, or by a combination of an antidepressant with another psychiatric drug.

For those who are helped by these medications, the relief may be so great that they are willing to endure their side effects and to take them indefinitely. Andrew Solomon's *New Yorker* article explains why he continues to rely on them:

> I hope not to have to go off my medications. I'm not addicted, because addicts are prone to symptoms caused by the removal of the drug, but I am dependent, because without the drug I would probably develop symptoms. I have some side effects which may eventually become intolerable. . . . The way that SSRIs undermine your capacity for sustained sexual fantasy means that you can climax only in the presence

of someone to whom you're strongly attracted. I gain weight more easily than I used to. I sweat more. My memory, which was never good, is impaired. I frequently forget in the middle of a sentence what I am saying. I get headaches often, and occasional muscle cramps. It's not ideal, but it seems to have put a real wall between me and depression.

The willingness of people to put up with the side effects of SSRIs has also encouraged their prescription for patients whose depression is much milder than Solomon's. Called dysthymic disorder—the first diagnosis I had assigned to Clara—these patients have Ham-D scores in the low to mid-teens. This is a far cry from the Ham-D scores of 25–30 found in patients with major depression. Nevertheless, symptoms of this magnitude are sufficiently distressing to lead them to seek help.

Studies of patients with dysthymic disorder show that they are indeed helped by antidepressants.[†] For example, in a double-blind study, more than half of the patients who had been treated for twelve weeks with imipramine or sertraline, an SSRI, had Ham-D scores of 4 or less, which was significantly better than the result achieved with a placebo. Nevertheless, as with major depression, some of the benefits are attributable to a placebo effect, and some of the patients got no benefit at all.

::

The successful development of drugs that relieve depression has come about despite a limited understanding of the way these drugs actually work. Figuring out the consequences of their effects on norepinephrine or serotonin has been particularly difficult because these neurotransmitters have such widespread effects throughout the brain. We now know that there are at least eight receptors for norepinephrine and fourteen receptors for serotonin, all of which are affected by prolonging the actions of their neurotransmitter. It is very surprising that such extensive drug-induced changes in neurotransmission have a therapeutic effect, instead of just disrupting many finely tuned brain functions.

The biggest mystery about antidepressants is that it generally takes weeks or more of continuous drug treatment for the therapeutic effect to develop.[†] Yet the influence of these drugs on neurotransmission is apparent within minutes after they are taken. This indicates that the immediate changes in neurotransmission are just the first step in a multistep process that relieves depression by gradually changing the brain.

Identifying the critical features of this gradual brain change has not been easy because the drugs affect so many aspects of brain physiology and chemistry. At first each drug elicits rapid countermeasures that are designed to bring the brain back to its unmedicated state. Such countermeasures regulate neurotransmission in the same way that a thermostat regulates the temperature of a room: when neurotransmission is augmented by a drug the brain thermostat turns on countermeasures to return it to its original level. Among these rapid countermeasures is a reduction in the release of neurotransmitters from nerve terminals and a reduction in the responsiveness of nerve cells to neurotransmitters.

But the brain's mechanisms of adaptation are much more complicated than those of the thermostat that controls room temperature. Over the course of several weeks of continuous drug treatment, the rapid countermeasures are gradually replaced by new modifications of the internal chemistry of the relevant nerve cells. Among the most important are modifications in the manufacture of two simple chemicals called cyclic AMP and cyclic GMP. These chemicals are generally referred to as "second messengers" because they carry signals that were initiated by the "first messenger"—the neurotransmitter—to various components of nerve cells, including their genes. By bringing signals to the genes, changes in the second messengers modify the production of hundreds of proteins in the affected nerve cells, thereby resetting their "thermostats" to a new position. It is this new state of cellular chemistry that is believed to be responsible for the antidepressant effect.[†]

Considering the complexity of this process, it is not surprising that its details have not yet been worked out. Many scientists have been examining antidepressant-induced changes in nerve cells in the hope of finding the main proteins that are affected by these drugs. Among the current favorites is a protein called BDNF[†] (brain-derived neurotrophic factor), which increases during sustained treatment with antidepressants. The increased BDNF influences many properties of nerve cells and their participation in brain circuits.

Antidepressants are not the only psychiatric drugs whose therapeutic effects develop slowly over the course of many weeks. The same is true of antipsychotic drugs[†] such as chlorpromazine. It is also true of the drugs that are used to treat a distinctive mood disorder called bipolar disorder.[†] Patients with this illness have episodes of manic grandiosity and excitement, as well as episodes of depression. Their mood swings can be controlled with a disparate group of drugs, called mood stabilizers, that includes lithium carbonate, carbamazepine (Tegretol), and valproate (Depakote).

Discovered by the same sorts of accidents[†] as the antipsychotics and antidepressants, the mood stabilizers have one critical difference: they do not immediately influence neurotransmission. Instead they work on the internal nerve cell machinery[†] to directly influence second messengers and the activities of several critical proteins. These changes, which occur over the course of several weeks, are believed to be responsible for the gradual stabilization of mood.

Despite the continued mystery about these gradual drug-induced changes, and despite the inconvenience of waiting for them to develop, the stable new states of the brain that they produce are ultimately helpful. Once they are established they can usually be maintained by continuing to take the drug. This progressive and sustained therapeutic effect of this group of drugs, which I call "delayed drugs," contrasts strikingly with the rapid therapeutic effects of another group of psychiatric drugs, which I call "immediate drugs." They are the subjects of the next two chapters.

4 :: Mother's Little Helper

It would be wrong and naive[†] to expect drugs to endow the mind with insights, philosophic wisdom, or creative power. These things cannot be provided by pills or injections. Drugs can, however, eliminate obstructions and blockages that impede the proper use of the brain. Tranquilizers, by attenuating the disruptive influence of anxiety on the mind, open the way to a better and more coordinated use of the existing gifts. By doing this, they are adding to happiness, human achievement, and the dignity of man.
—Frank M. Berger (1970)

Mother's little helpers were pills—Miltown, amphetamine, barbiturates, Librium, and Valium were the most popular and widely available in the fifties and early sixties—that were used to keep women in their place, to make them comfortable in a setting that should have been uncomfortable, to encourage them to focus on tasks that did not matter. I cannot think of the phrase even today without hearing it in Mick Jagger's sneering tones.
—Peter D. Kramer (1993)

The discoveries of chlorpromazine and imipramine in the 1950s were landmark events not only because of the relief they provided to so many people, but also because they encouraged a search for other types of psychiatric drugs. If simple chemicals could eliminate the bizarre ideas of schizophrenia or lift the black cloud of depression, it seemed reasonable to look for pharmacological remedies for other forms of mental distress.

Of these the most prevalent is the uncomfortable state of heightened awareness that we call anxiety. Under appropriate circumstances, this feeling of distress is a useful warning of impending danger, and it subsides when the danger has passed. But many people become anxious in the absence of substantial hazards. Such abnormal anxiety takes different forms, which are the defining characteristics of particular anxiety disorders. Each is an exaggerated version of a normal response to certain threatening situations.

The main feature of the most dramatic of these disorders, called panic disorder, is recurrent unexpected attacks of extreme fear that come on for no apparent reason. During these panic attacks, which usually last for just a few agonizing minutes, there are extreme physiological changes such as rapid breathing and a pounding heart that would be appropriate only in response to the greatest authentic danger. Along with this overwhelming and inexplicable distress is the feeling of losing control, and a persistent concern that another attack may start at any time.

Although panic attacks often come out of the blue, many patients are more likely to get them in places or situations where they feel particularly vulnerable. This feature is called agoraphobia (from the Greek for "fear of the agora," or marketplace) because the typical situation that is feared is leaving home for a public place. Some people also have more specific fears, such as being in a crowd or traveling in a train. Avoidance of these situations may become disabling and lead to a self-imposed house arrest.

In another form of anxiety disorder the discomfort is less intense but more sustained. Called generalized anxiety disorder, its main feature is persistent inappropriate worries that bad things will happen. People with this disorder may be preoccupied for years with concerns that misfortune will strike them or their children, and that they will lose their money or their job. Along with these apprehensive expectations there is usually irritability and difficulty sleeping— the sort of disquietude that accompanies the realistic and transient worries that we all experience. Patients with generalized anxiety disorder frequently seek help for physical complaints and are among the largest consumers of medical services.

Other forms of anxiety disorder are largely restricted to a specific situation. The most prevalent is social phobia. Also called social anxiety disorder, its essential feature is a marked and persistent fear of social situations or of exposure to an audience, as in public speaking. The main concern of people with social phobia is that they will be humiliated by displaying their incompetence and lack of self-control while undergoing the scrutiny of others. Although the sufferers know full well that their anxiety is excessive or unreasonable, they avoid the feared situation or endure it only with an inner feeling of intense distress.

Obsessive-compulsive disorder is also driven by unreasonable anxiety. In typical cases the anxious feelings take the form of obsessive thoughts, such as the fear of becoming contaminated by germs, and some relief comes by performing a compulsive ritual, such as repeated hand washing. But the relief is transient, and the thoughts and rituals are repeated again and again.

At least one person in ten people suffers from one of these anxiety disorders, and many methods have been developed to treat them. The first was invented about a hundred years ago by Sigmund Freud, who believed that anxiety is the basis of all mental disorders. Called psychoanalysis, this method aims to relieve anxiety, in all its forms, by attempting to discover its hidden origins in unconscious conflicts between desires and the fear of punishment. Newer psychotherapies for anxiety, such as cognitive therapy and behavioral therapy,[†] pay little attention to origins. They are mainly concerned with teaching people how to either avoid situations that elicit irrational fears or develop techniques for confronting them and diminishing their intensity. Many sufferers from anxiety benefit from psychological treatment. But others reject the slow relief that it promises. Instead they choose the more rapid relief that they can get from a pill.

::

During the first half of the twentieth century the most widely prescribed drug for anxiety was phenobarbital. But the relief it provided was achieved at the expense of considerable sleepiness and dulling of wits. Because of these and other drawbacks, physicians were on the lookout for a better alternative.

The drug that first replaced phenobarbital as a treatment for anxiety grew out of unexpected observations about a chemical that kills bacteria. The observations were made by Frank Berger, a Czech physician who had fled to England to escape the Nazis and took a job as a laboratory scientist with the British Drug House. While testing the antibacterial chemical in mice, Berger noticed that it was a powerful muscle relaxant. After emigrating to the United States, Berger encouraged a chemist at Carter Products, a small pharmaceutical company, to make derivatives of this interesting chemical—the same sort of advice that led Charpentier to make chlorpromazine. In May 1950 the chemist, B. J. Ludwig, created the derivative that became known as meprobamate.[†]

Were it not for the discovery of the benefits of chlorpromazine, Ludwig's creation might have languished in the New Jersey storehouse of Wallace Laboratories, Carter's subsidiary. But the rapid acceptance of Rhône-Poulenc's new medicine encouraged Berger to find out if meprobamate could also be used for some form of mental distress. What particularly excited him was his observation that meprobamate seemed to be better than phenobarbital in calming laboratory monkeys. Convinced of its tranquilizing effects, Berger persuaded Wallace Laboratories to try to market the drug as a new treatment for anxiety, under the trade name of Miltown. He also hedged his bets by licensing it to Wyeth, a Philadelphia drug company, which sold the identical drug under the trade name of Equanil.

Initial clinical studies of meprobamate showed that it helped anxious patients. Although none of these was a randomized controlled study, and none provided rigorous evidence of effectiveness, they met the much lower standards of the time. In one study of almost two-hundred patients, published in 1955 in the widely read *Journal of the American Medical Association*, L. S. Selling reported that "Miltown was of considerable value[†] in anxiety and tension states" and that of the patients who had previously taken phenobarbital, "none of these patients preferred phenobarbital to Miltown."

But these conclusions were not based on double-blind comparisons with placebos.

Nevertheless, physicians quickly accepted Miltown as a breakthrough drug that might even be as important as chlorpromazine. In this view chlorpromazine was the prototype of a category of drugs called major tranquilizers, for severe mental disorders, whereas Miltown was the prototype of minor tranquilizers, for milder forms of mental distress. In the late 1950s Miltown became the best-selling drug in America.

This enthusiastic acceptance changed the public's attitude toward the use of psychiatric drugs. Before Miltown it was widely believed that drugs should be prescribed only for disabling illnesses such as schizophrenia, and that milder forms of mental distress should be handled with psychological treatments or self-discipline. But the widespread prescription of Miltown convinced the public that psychiatric drugs should be used more freely.

It was Berger's zeal for this drug that played a large part in its early success. To him its greatest value was in dampening useless emotions. In a paper describing the development of Miltown[†] he begins with a citation from the seventeenth-century philosopher Spinoza: "Human bondage consists in the impotence of man to contain the affects . . . for a man who is under their control is not his own master. A free man is one who lives according to the dictates of reason alone." In Berger's view Miltown was not a specific treatment for a particular category of mental disorder but, instead, a general remedy that helped people to overcome the tyranny of the affect that we call anxiety, so that reason could prevail. Unlike phenobarbital, which he dismissed as an old-fashioned sleeping pill, Miltown was promoted as a modern remedy that could bring about optimal functioning and happiness.

But despite Berger's fervent advocacy and the enthusiastic reports of many physicians and patients, it became clear by the early 1970s that Miltown wasn't really much better than phenobarbital after all.[†] Like phenobarbital, Miltown reduced anxiety only at the price of some sleepiness, was lethal if taken in large doses, and was addictive. As David Greenblatt and Richard Shader put it in a scathing 1971 review: "meprobamate, although pharmacologically distinct from the barbiturates, is no less toxic and no more effective in reducing anxiety than a barbiturate such as phenobarbital."

The review by Greenblatt and Shader in the *American Journal of Psychiatry*, entitled "Meprobamate: a study of irrational drug use," went on to challenge the claim that Miltown worked at all. In their summary the authors conclude that "the history of the tranquilizer meprobamate illustrates how factors other than scientific evidence may determine patterns of physicians' drug use. Forceful advertising and publicity, an attitude of general optimism, and uncontrolled studies with favorable results [such as Selling's 1955 report] combined to elevate meprobamate to the position of America's magical cure-all tranquilizer. This drug remains in wide use despite a large body of sound scientific data that questions its efficacy."

The rise and fall of Miltown each had lasting effects on psychiatry. Belated questions about its effectiveness contributed to the establishment of the requirement that psychiatric drugs could be brought to market only after their value was established by means of randomized placebo-controlled trials. On the other hand, the eagerness with which Miltown was initially received encouraged the development of a new group of medicines that are truly better treatments for anxiety.

::

The progenitor of these truly better drugs was already sitting in a test tube in the Nutley, New Jersey, laboratories of Hoffman–La Roche in 1955, the year that Miltown came to market. The product of tinkering by Leo Sternbach,[†] another immigrant from Eastern Europe, it had been created in the hopes of finding a tranquilizing drug like chlorpromazine. The only reason Sternbach made this particular chemical is that he had prior experience with similar ones. Having no reason to believe that it would have effects on the brain, he left it lying around on his cluttered laboratory bench until 1957. Then, in the process of spring cleaning, he sent it off to his colleague, Lowell Randall, for testing in mice, as compound R050690.

Like chlorpromazine, Ro50690 had obvious effects on the behavior of laboratory rats. Within a few days of receiving it from Sternbach, Randall informed him that it had a sedative effect if given in large doses. Even more impressive was the subsequent finding that smaller doses reduced anxiety without much sedation. For example, when mice treated with Ro50690 were presented with a threatening signal, such as a tone that is sometimes followed by a shock, they seemed less perturbed than nonmedicated animals. Nevertheless, the drug-treated animals were still capable of efficiently performing tasks that require alertness, showing that they were not very sedated. In later studies it was shown that the medicated animals actually perform better on tests conducted right after a threatening signal, because they are not distracted by intense anxiety.

Encouraged by the studies in experimental animals that suggested that this new drug was a potential competitor of Miltown, Hoffman–La Roche asked a few psychiatrists to test it on patients in early 1958. They found that it was a very good treatment for anxiety. Although these initial studies were no more rigorous than those that had been done with meprobamate, randomized placebo-controlled trials subsequently confirmed the new drug's effectiveness. In 1960 it was approved for sale under the generic name of chlordiazepoxide and the trade name of Librium.

Over the next few years Librium became even more popular than Miltown. Soon it was joined by another member of this class of drugs, called benzodiazepines, which Sternbach made by modifying Librium. Named diazepam and marketed as Valium, this drug is about five times as potent as Librium. Valium also starts working faster than Librium because the drug is absorbed more rapidly from the intestines. Heavily promoted by Roche as a superior product, Valium—Mick Jagger's " little yellow pill"—soon replaced Librium at the top of the best-seller list, and remained there through most of the 1970s.

Stimulated by the enormous success of Valium, several pharmaceutical companies introduced other benzodiazepines. The main difference among them is in their duration of action, which is determined by the speed with which the body converts them into inactive chemicals. Short-acting benzodiazepines such as triazolam (Halcion) were designed to be used as sleeping pills, because their effects will have worn off by the morning. Longer-acting ones such as clonazepam (Klonopin) and an extended-release form of alprazolam (Xanax XR) are preferred for sustained relief of anxiety.

After years of experience it is generally agreed that benzodiazepines are better drugs than phenobarbital and meprobamate. First, they relieve anxiety at doses that are somewhat less sedating. Second, they are much safer because overdoses are not lethal—except in combination with large amounts of other drugs, such as alcohol. As a result, benzodiazepines have largely replaced their predecessors as treatments for anxiety.

::

When Librium and Valium became the favorite new treatments for anxiety in the 1960s, their effects on the brain were still as mysterious as those of phenobarbital and meprobamate. The mystery remained until 1975, when the first major clue was discovered by Erminio Costa—another of the gifted disciples of Bernard Brodie—and by Willy Haefely, a pharmacologist at Hoffman–La Roche. Both Costa and Haefely showed that the benzodiazepines work by augmenting the actions of a neurotransmitter called gamma-aminobutyric acid.[†]

Generally referred to as GABA, this neurotransmitter is made in the brain by a simple modification of glutamate that radically alters its function. Whereas glutamate is the main activator of brain circuits, GABA is their main inhibitor. Its action as an inhibitor also distinguishes GABA from the "big three" amines—dopamine, norepinephrine, and serotonin—which generally stimulate brain circuits. If the big three are viewed as specific accelerators that keep certain brain machines going, GABA is the overall brake.

Another difference is that GABA, like glutamate, is used as a neurotransmitter by a large fraction of the nerve cells in the brain, whereas dopamine, norepinephrine, and serotonin are each concentrated in a small number of nerve cells. But despite its widespread distribution, GABA has selective effects on particular brain circuits. Some of the nerve cells that use it as a neurotransmitter act as brakes on the circuits that control anxiety. Others act as brakes on circuits that wake up the brain, so that augmentation of the action of GABA promotes sleep.

To further complicate matters, there are many different types of GABA receptors, each made up of various combinations of different protein building blocks that are arranged together in groupings of five. Furthermore, each of these groupings has a different distribution in the brain. Benzodiazepines augment the actions of GABA by binding to a specific site (called the benzodiazepine binding site) on particular building blocks of GABA receptors, thereby changing the shape of these receptors in ways that make the GABA signal last longer—a way of sustaining the application of the brakes.

Benzodiazepines are not the only drugs that augment the actions of GABA. Phenobarbital, meprobamate, and alcohol also change the shapes of certain GABA receptors[†] and prolong certain GABA signals, and this accounts for their sedative and antianxiety effects. But none of those drugs binds to the benzodiazepine binding site,[†] and none produces exactly the same changes as benzodiazepines. For example, benzodiazepines have different effects on GABA receptors that control breathing than the other drugs. This explains why overdoses of phenobarbital, meprobamate, or alcohol can produce a fatal cessation of breathing, whereas overdoses of benzodiazepines do not.

Despite these selective effects, all these drugs have an important feature in common: if taken for prolonged periods (usually weeks or more) in large enough doses, each of them may begin to lose its effectiveness. Called tolerance, this loss of effectiveness reflects countermeasures that the brain mounts to overcome the effects of the drug[†] and return to its unmedicated state. To reinstitute the drug effect, it is necessary to increase the dose. Furthermore, because of these countermeasures, the brain may become dependent on the continued presence of the drug. Should the drug be abruptly stopped, withdrawal symptoms (increased anxiety and irritability) can result.

When benzodiazepines were first introduced and were pre-scribed in large doses, they often produced extreme tolerance and severe withdrawal symptoms. To prevent such profligate use, the U.S. government mandated safeguards against their excessive pre-scription. But the relatively small doses of benzodiazepines that are presently used to treat anxiety rarely lead to the craving and com-pulsive out-of-control use that are the hallmarks of addiction.

Nevertheless, the potential dangers of tolerance, dependence, withdrawal, and addiction have made some psychiatrists reluctant to prescribe benzodiazepines, except for short periods of time. Their reluctance has been intensified by the discovery of an alternative group of medications for anxiety disorders that don't have this draw-back. As with many other discoveries about psychiatric drugs, this one came as a result of tinkering.

::

The tinkerers in this case were Donald Klein and Max Fink, two psychiatrists who were curious about the proper uses of imipramine, which Geigy began marketing in Europe in 1957. Although Kuhn had already shown that imipramine relieves major depression, Klein and Fink wondered how it would affect other types of patients. In October 1958 they began a radical experiment:[†] they gave imipra-mine to just about everyone on the psychiatric ward of Hillside Hos-pital on Long Island, New York. Of the roughly two hundred patients who received at least three weeks of treatment with this drug in the course of this study, most had the diagnosis of schizo-phrenia or a mood disorder.

To Klein and Fink these diagnoses didn't matter. Instead of clas-sifying each patient on the basis of a pattern of symptoms, as defined in a diagnostic manual, they decided to classify them on the basis of their response to imipramine. Using this unorthodox approach, they confirmed some of the earlier observations about the drug, such as its antidepressant effect.

But there was also a surprise. Fourteen of the patients who received the drug had an unanticipated response: reduction of episodic anxiety. These patients reported that they had been suffering attacks "accompanied by rapid breathing, palpitations, weakness and a feeling of impending death," which we would now consider to be indications of panic disorder. Their considerable improvement after weeks of treatment with imipramine called attention to this distinctive pattern of symptoms, and helped to define panic disorder as a diagnostic category.

The discovery that sustained treatment with imipramine can prevent panic attacks, which was confirmed by placebo-controlled studies, led to a rethinking of the classification of this psychiatric drug. It gained urgency when other studies showed that prolonged treatment with MAO inhibitors also prevented panic attacks. So why call imipramine and MAO inhibitors antidepressants[†] when they are also effective remedies for a severe anxiety disorder?

Rethinking the classification of imipramine and MAO inhibitors also led to a rethinking of the norepinephrine hypothesis of depression, which was widely accepted at that time. Given that norepinephrine has an activating effect on the brain, it made sense that imipramine and MAO inhibitors—both of which augment the actions of this neurotransmitter—would energize brain functions and counteract the lethargy of depression. But increasing brain activation with these drugs should make panic worse, instead of better—just the opposite of the effects that were observed.

Lack of understanding of the way these drugs work did not, however, stand in the way of their being prescribed for patients. Having decided that "the drug reaction was the determining feature" and that they would not be constrained by "the standard diagnostic nomenclature," Klein and Fink had made an important practical discovery. By the 1980s imipramine was being widely used for panic disorder.

::

It was only natural, therefore, to test the effects of every new antidepressant on patients with panic disorder. In the course of the 1990s all of the popular serotonin reuptake inhibitors—fluoxetine (Prozac), sertraline (Zoloft), paroxetine (Paxil), fluvoxamine (Luvox), and citalopram (Celexa)—have been successfully used to prevent panic attacks. In "Practice Guidelines for the Treatment of Panic

Disorder," a booklet published in 1998 by the American Psychiatric Association, the SSRIs are considered to be interchangeable.[†] They are now prescribed for panic disorder much more frequently than imipramine, because most patients are less troubled by their side effects.

The great acceptability of SSRIs has also encouraged their testing on patients with other common anxiety disorders. Placebo-controlled studies with some members of this class of drugs have established their effectiveness for many forms of pathological anxiety. Another drug that influences serotonin neurotransmission, called buspirone (BuSpar), is also used for this purpose.

The fact that SSRIs can alleviate such a wide range of symptoms makes them especially popular with psychiatrists, who are happy to be able to deal with multiple problems with one pharmacological blow. This was the situation I inadvertently found myself in with Clara, whose depression and obsessive preoccupation with her nose were both relieved by Prozac. For the many patients, such as Andrew Solomon, whose symptoms justify the diagnosis of both a mood disorder and an anxiety disorder,[†] it is nice to be able to offer such versatile drugs.

Recognizing the appeal of multipurpose medications, pharmaceutical companies often emphasize this feature of their wares. For example, Pfizer's famous ad for Zoloft shows an ecstatic man successfully tossing a paper ball into a wastebasket with the caption "A Three-Pointer," exulting in the good fortune that this same medication can relieve the symptoms of major depression, panic disorder, and obsessive-compulsive disorder. Just as dermatologists have learned to use hydrocortisone ointment for many kinds of rashes, so too have psychiatrists learned to use an SSRI for many kinds of mental distress.

But despite their value in treating anxiety disorders, SSRIs have not put benzodiazepines out of business. Unlike the SSRIs, which are delayed drugs that don't relieve symptoms for many weeks, benzodiazepines are immediate drugs, which start working right away. For this reason the initial treatment of patients with anxiety disorders often includes a combination of a benzodiazepine and an SSRI as well as cognitive and behavioral psychotherapy. The benzodiazepine brings prompt relief. By lowering anxiety, it may facilitate the psychological treatment, which teaches the patient how to deal with situations that provoke symptoms, and how to manage them when they appear. As the psychotherapy and the SSRI take effect in the ensuing weeks, the benzodiazepine is gradually withdrawn. This greatly reduces the risk of abuse of the benzodiazepine, which may occur if it is taken for longer periods.

Benzodiazepines are not the only widely used psychiatric drugs that are sometimes abused. They share this danger with another group of immediate drugs, to which I now turn.

5 :: *A Spectacular Improvement*

The psychological reactions of 30 behavior problem children who received benzedrine sulfate for one week were observed. There was a spectacular improvement in school performance in half the children.
—Charles Bradley (1937)

In 1937 Charles Bradley, a child psychiatrist, conducted an experiment with a drug that was then widely used as a nasal decongestant. Having learned that adults became more alert after taking this medication, Bradley wondered how it might affect children with mental disorders. He decided to do his experiment because "the psychological response of children to drug therapy[†] of any sort appears to have been a generally neglected subject." The results put an end to this neglect. They eventually led to the active use of psychiatric drugs for certain patterns of childhood behavior—a practice that some now regard as excessive.

As subjects for his experiment, Bradley selected thirty children who were hospitalized at the Emma Pendleton Bradley Home, a hospital for psychiatric disorders in East Providence, Rhode Island. The group was made up of twenty-one boys and nine girls, which reflected the sex ratio of his patients. They were up to twelve years old and had already been in the hospital for at least a month. Bradley made no attempt to choose patients with a particular type of behavioral abnormality or a particular diagnosis. His only criterion was that their "behavior disorders were severe enough to have warranted hospitalization."

Bradley's study of the effects of the drug, called Benzedrine, was conducted in three weeks. In the first week the patients were carefully observed by the nurses and teachers who were already assigned to them, and received no medication. During the second week, each patient received a Benzedrine pill upon rising, and was observed as before. Observations continued in the third week, but no further drug was given. When these observations were completed, the records of behavior during the three weeks were compared.

Bradley was startled by the results. As he explained:

Possibly the most striking change in behavior[†] during the week of benzedrine therapy occurred in the school activities of many of these patients. Fourteen children responded in a spectacular fashion. Different teachers, reporting on these patients, who varied in age and school accomplishment, agreed that a great increase of interest in school material was noted immediately. . . . Speed of comprehension and accuracy of performance were increased in most cases. . . . The improvement was noted in all school subjects. It appeared promptly the first day benzedrine was given and disappeared on the first day it was discontinued.

Bradley went on to comment:

To see a single daily dose of benzedrine produce a greater improvement in school performance than the combined efforts of a capable staff working in a most favorable setting, would have been all but demoralizing to the teachers, had not the improvement been so gratifying from a practical viewpoint.

::

The Benzedrine that Bradley used for this influential experiment had been given this name by Gordon Alles, the chemist who had synthesized it[†] in 1927, and who first called attention to its physiological and psychological effects. The name came from one of the chemical appellations of the compound—benzyl-methyl carbinamine. Benzedrine became its trademark when it was introduced in 1932 by Smith, Kline, & French, the company that would later bring chlorpromazine to market in the United States. It is now better known by the generic name it was given by the Council on Pharmacy and Chemistry of the American Medical Association. Deriving a name from another chemical appellation, alpha-methyl-phenylethylamine, they called it amphetamine.

Alles made amphetamine while working in a small laboratory in the office of an allergist in Los Angeles. His main job was to purify proteins from substances that triggered asthma and other allergic reactions so that they could be used for a treatment called allergic desensitization. In his spare time he set out to make a new drug to relieve the symptoms of asthma.

Alles' approach to making this new drug derived from the work of K. K. Chen, a pharmacologist at the University of Wisconsin in Madison. Chen had become interested in ma huang, a Chinese herbal remedy that is obtained from a shrub called *Ephedra vulgaris* and which had been used for more than five thousand years as a decongestant. The active ingredients of ma huang had already been identified by the end of the nineteenth century by Japanese and German scientists, who named them ephedrine and pseudoephedrine. Chen confirmed the presence of these compounds in ma huang in a sample he obtained in Beijing. Then, in 1926, he described a novel method for making ephedrine from simple chemicals—a method he considered so important that it "marks one of the triumphs attained in the field of synthetic chemistry."[†]

Like the ephedrine extracted from ma huang, the synthetic compound proved to be very useful for treating asthmatic attacks, by relaxing the spasm of the small tubes in the lung called bronchioles. In this it mimics the action of adrenaline, which had already been isolated from adrenal glands at the end of the nineteenth century and was known to be useful for treating asthma. The great advantage of ephedrine is its effectiveness if taken as a pill, whereas adrenaline works only if injected or inhaled. Furthermore, ephedrine lasts much longer than adrenaline, which is rapidly destroyed by the body. A single pill of ephedrine may terminate an attack of asthma, whereas repeated doses of adrenaline may be required to do this job. Ephedrine's only drawback was that, like many drugs, it was very expensive when it was first brought to market.

Alles set out to find a cheaper substitute. Knowing the structures of both adrenaline and ephedrine, he soon found an economical way to make the related compound that he named Benzedrine. But much to Alles' disappointment, Benzedrine doesn't work for asthma. It does, however, share ephedrine's ability to relieve nasal congestion. It also has another useful property: when prepared in a particular chemical form, it gives off vapors that can be inhaled. Packaged as the Benzedrine Inhaler, a little plastic tube you can keep in your pocket, it became an extremely popular remedy for stuffy noses, because a few inhalations through the nose gave rapid relief.

In the course of his studies of Benzedrine, Alles tried it himself. He noticed that a single pill made him more alert for several hours. Unlike ephedrine, which made him jittery—simulating the panicky feeling that can come from injections of adrenaline—small amounts of Benzedrine produced a pleasant form of alertness that was different from the heightened arousal that accompanies fear.

Alles' observations about the activating effects of Benzedrine led several of his colleagues to use it for the treatment of narcolepsy.[†] People with this disease, which affects about one person in two thousand, repeatedly fall asleep for short periods throughout the day. They also have attacks of limpness, called cataplexy, which may be precipitated by strong emotions such as anger or laughter. This was taken as evidence that narcolepsy is an emotional disorder, and led to attempts to treat it with psychotherapy, which proved to be ineffective. In contrast, Benzedrine provided considerable relief.

Alles and his colleagues also discovered that only half of the Benzedrine was responsible for these effects. Like many other chemicals, Benzedrine (amphetamine) exists in a right-handed form (dextro-) and a left-handed form (levo-), and the two forms have different properties. In the case of amphetamine it is the right-handed form, named dextroamphetamine, that is activating. For this reason,

Smith, Kline, & French prepared pills that contained only dextroamphetamine. The trademarked preparation, called Dexedrine, is still used to treat narcolepsy.

The stimulant effects also suggested that amphetamine might be used to treat depression. In 1936 Abraham Myerson, a psychiatrist at Boston State Hospital, published a paper entitled "Effect of benzedrine sulfate on mood and fatigue in normal and neurotic persons," which encouraged its use for a variety of patients. Myerson concluded that the drug—which he gave in the morning so that it would not interfere with sleep—"seems to have a definite though limited value in combating the neuroses.[†] It can be definitely stated to be an ameliorative agent rather than a curative one; that is, when used judiciously it is of value in lessening the distress and the depression and increasing the feeling of energy." It was Myerson's report that had caught the eye of Charles Bradley and led him to conduct his influential experiment on children.

::

The main effect of amphetamine that Bradley observed in his hospitalized children was not what he had anticipated. Unlike the adults observed by Myerson and others, who were activated by the drug, many of the children Bradley studied became what he called "subdued." As Bradley put it in his 1937 report: "Fifteen of the 30 children responded to benzedrine by becoming distinctly subdued in their emotional responses. Clinically in all cases this was an improvement from the social viewpoint. . . . In this group were some children who had expressed their irritability in group activities by noisy, aggressive, domineering behavior. Such children under the influence of the drug became more placid and easy-going."

In a subsequent study—"Amphetamine (benzedrine) therapy of children's behavior disorders," which he published in 1941 with his colleague, Margaret Bowen—Bradley described the results of amphetamine treatment of another one hundred children, ages five to twelve, who were resident patients at his hospital. Again the children had a variety of disorders, and again the main effect, which he observed in more than half of them, was that they were subdued.

Bradley and Bowen elaborated on the response they observed:

By a subdued response is meant[†] that in some conspicuous way a child becomes less active than before. The term is employed in a social rather than a physiological sense. Many children began to walk and move quietly in contrast to previous noisy running and rushing about. A number spoke in a normal or lowered voice instead of shouting raucously. Some of these same children, instead of quarreling and arguing boisterously, began to avoid expressing differences of opinion or conducted their discussions in tones which were not offensive. In certain instances children appeared subdued because they began to spend their leisure time playing quietly or reading, whereas formerly they wandered aimlessly about antagonizing and annoying others.

The general impression given by all the children who became definitely subdued was that they were effectively exerting more conscious control over their activities and the expression of their emotions. In general, they were conducting themselves with increased consideration and regard for the feelings of those about them. Mental alertness was always present. There was no evidence of drowsiness, sluggishness, inertia, or any of the retarded responses and the intellectual confusion sometimes noted following therapeutically effective doses of narcotic or sedative drugs.

But not all of the patients were subdued. Nineteen of them were "generally stimulated by the drug. Such children became more alert, accomplished their daily tasks with more initiative and dispatch, became more aggressive in competitive activities, and showed an increased interest in what was going on about them." In contrast with the fifty-four subdued children, who were all judged to be "improved," only twelve of the nineteen stimulated children were "improved," while seven were "worse." Six children who were neither stimulated nor subdued showed improved school performance, whereas the remaining twenty-one showed no effect at all. Bradley and Bowen detected no relationship between the response of the children to the drug and their clinical diagnosis.

Gratified by the value of amphetamine for so many of their patients, Bradley and Bowen were at a loss to explain its effects. They were particularly puzzled by two questions: Why does a drug that activates adults subdue so many young patients? Why do some children benefit from being subdued by the drug whereas others benefit from being stimulated by it?

::

These questions, which have still not been satisfactorily answered, were entirely based on the behavioral changes that Bradley and Bowen observed. None of the effects of the drug on brain chemistry had been discovered when they wrote the paper I just cited. In fact, dopamine, the main neurotransmitter that amphetamine affects, had not even been detected in the brain in 1941.

In the years since Bradley's initial discovery, amphetamine's effects on the brain have been studied in great detail. Like many other psychiatric drugs, amphetamine has multiple effects on chemical neurotransmission. These result from its binding to the transporters on nerve terminals that control the reuptake of norepinephrine, serotonin, and—most importantly—dopamine. When amphetamine gets into the brain, it binds to the dopamine transporter, alters its structure, and reverses the direction of flow of the neurotransmitter. Instead of pumping dopamine back into the nerve terminal, the altered transporter pumps dopamine out into the synaptic cleft. The large amount of dopamine in the synaptic cleft activates the adjacent nerve cell. It, in turn, activates the other nerve cells in the brain circuits in which it participates.

The behavioral effect of amphetamine depends on the dose of the drug and the way it is taken. In Bradley's experiment he gave small doses in the form of pills. Once an amphetamine pill is swallowed, the drug is gradually absorbed from the intestinal tract, enters the bloodstream, and is transported to the brain. There it activates brain circuits that produce the rapid changes in behavior characteristic of an immediate drug.

The single daily dose of amphetamine that Bradley administered is sufficient to activate these brain circuits for only a few hours. As the drug is destroyed in the body, the transporters start working normally again, dopamine is pumped back into the nerve terminals, and the brain circuits it activated return to their previous state. So too do the nerve terminals that were partially depleted of dopamine because of the actions of the drug. The next daily dose has the same effect as the prior one.

If amphetamine is taken in larger doses by injection, the effects are very different. Instead of the trickle of drug into the brain after swallowing a pill, there is a sudden torrent of the drug within seconds after it is delivered into the bloodstream. This leads to a flooding of synaptic clefts with dopamine and intense activation of the circuits that dopamine controls. The main experience of this surge of dopamine is a transient feeling of intense pleasure that is largely due to the action of this neurotransmitter in a tiny region of the brain called the nucleus accumbens. As the pleasure subsides it is followed by fatigue and depression, which may persist until the depleted stores of dopamine are replaced and the exhausted nerve cells recover.

Other changes in the brain may persist long after the amphetamine is gone. The most striking is an increased sensitivity to certain actions of the drug, called sensitization.[†] This and other changes that develop with prolonged use of this immediate drug have very different consequences than those that are set in motion by the delayed drugs such as Prozac and chlorpromazine. In the case of the delayed drugs the changes in the brain are responsible for their gradual therapeutic effect. In contrast, the changes in the brain that develop with frequent large doses of amphetamine or benzodiazepines are hardly therapeutic. With both of these categories of immediate drugs gradual modifications in brain chemistry and physiology may lead to their compulsive use—the well-known pattern of behavior called addiction.[†]

Some Features of Immediate and Delayed Drugs

	Immediate drugs	Delayed drugs
Examples	Diazepam (Valium) Amphetamine	Chlorpromazine Fluoxetine (Prozac)
Therapeutic onset	Immediate	Weeks (but some immediate)[1]
Slow brain reactions	Undesirable	Therapeutic
Addictive	Yes	No
Stop gradually[2]	Yes	Yes

[1] Delayed drugs have some immediate effects. Many of them are side effects, but some are therapeutic. For example, chlorpromazine can quickly terminate a manic episode, and Prozac can quickly relieve the irritability and depression that may be associated with the premenstrual period.
[2] Because both delayed and immediate drugs produce gradual changes in the brain, they should all be stopped gradually, by progressive reduction of the daily dose. This allows the brain to slowly adjust to their absence, and avoids symptoms that may accompany their abrupt withdrawal.

The addictive potential of intravenous amphetamine first became apparent in the 1950s. Until then its many effects on the brain led to the widespread prescription of amphetamine pills for a variety of purposes. For example, amphetamine's inhibition of appetite, through actions on serotonin, made it popular with dieters. Its activating effect, through actions on norepinephrine, made it a favorite of students studying for exams, as well as soldiers in battle conditions. Millions of people used amphetamine pills for these and other purposes without becoming addicted. It was only when people began injecting the drug that it became apparent that it was just as addictive as intravenous cocaine, which also produces a flood of dopamine by binding to the dopamine transporter. And like intravenous cocaine, intravenous amphetamine can produce transient paranoia, which, as I have pointed out, resembles that seen in schizophrenia.

Because amphetamine has such a great potential for abuse, its prescription is controlled by U.S. federal laws, using criteria established by the Comprehensive Drug Abuse Prevention and Control Act of 1970. This act, whose details were more influenced by politics than by science, classifies drugs that are prone to abuse on a five-point Roman numeral scale, where I is the most dangerous and V is the least dangerous. Schedule I drugs include heroin and LSD, which are considered to have no justifiable medical use and can be used only for experimental purposes licensed by the federal government. Amphetamine is classified as a Schedule II drug, along with morphine and certain barbiturates. Schedule II drugs are considered to be as dangerous as Schedule I drugs but have clear medical benefits, and their prescription is carefully regulated. This distinguishes amphetamine from benzodiazepines such as Valium, which are classified as Schedule IV drugs to indicate that they are less likely to be abused. But despite the great dangers of large doses of amphetamine, especially if taken intravenously, small daily oral doses of amphetamine are now used safely by millions of children—because of the effects that were discovered by Charles Bradley.

::

The behavioral effects of amphetamine that Bradley observed are not confined to hospitalized children with severe mental disorders like those that he studied. Other children who receive amphetamine also perform better in school. This is especially true of a group of children who are much more active and impulsive than their peers and who are so distractible that they may pay little attention to their parents and teachers. In Bradley's time children with extreme forms of this pattern of behavior were said to have "minimal brain dysfunction."[†] Now their official diagnosis is attention deficit hyperactivity disorder (ADHD). Because hyperactivity is not always present, it is sometimes referred to as ADD.

The essential features of ADHD that are enumerated in psychiatry's diagnostic manual, *DSM-IV,* include a persistent pattern of inattention, hyperactivity, and impulsivity "that is more frequent and severe than is typically observed in individuals at a comparable level of development." As with many other psychiatric diagnoses, the defining features of abnormality are a matter of degree. Although the current diagnostic scheme is categorical—a yes-or-no scheme in which a patient either has enough characteristics to fit into a particular category or doesn't—many psychiatrists actually use a graded or dimensional diagnostic scheme for certain disorders[†] such as ADHD. To define the categorical diagnosis of ADHD *DSM-IV* requires "clear evidence of interference with developmentally appropriate social, academic or occupational functioning"—but this, too, can be thought of in a graded rather than a yes or no way. To further complicate matters, some child psychiatrists have pointed out that the very same distractibility and hyperactivity that can interfere with school performance may also be correlated with desirable attributes[†] such as creativity and adventurousness. Nevertheless, using the set of criteria that define the categorical diagnosis of this disorder, *DSM-IV* estimates its prevalence as "3–5 percent of school-age children," most of them boys.

A major reason that there is such interest in making this diagnosis is that drugs can modify the undesirable behavior. Double-blind tests have repeatedly shown that amphetamine improves the general conduct and school performance of many of the children who fit this profile. So too does treatment with a related drug called methylphenidate (Ritalin), whose stimulant properties and resemblance to amphetamine were discovered in 1954 by scientists at Ciba, a Swiss drug company. Introduced in the United States in 1955 as a treatment for narcolepsy, it was approved in 1970 for the treatment of children with "minimal brain dysfunction," the early term for ADHD.

Like amphetamine, methylphenidate binds to transporters that control the uptake of brain amines, especially the dopamine transporter. And, like amphetamine, intravenous methylphenidate produces a surge of brain dopamine that can be so addictive that it is classified as a Schedule II drug. Both drugs can change the behavior of children with ADHD, although one may work better than the other for an individual patient.

A major drawback of standard amphetamine or methylphenidate pills is that they last for only a few hours, so schoolchildren who are given a dose in the morning may need a second one at lunchtime. But both drugs are now available as pills that release their contents slowly, so that a single morning dose lasts throughout the school day. For example, amphetamine is now available in a long-lasting proprietary form called Adderall-XR. Novel types of pills that deliver methylphenidate for twelve hours were introduced several years ago under the trade names Concerta and Metadate-CD. The latter is being advertised by a Superman-type cartoon character who exclaims, "Kids, I'll be there when you need me." Not to be outdone, the manufacturer of Ritalin has introduced a long-acting version (Ritalin LA).

Like all psychiatric drugs, amphetamine and methylphenidate have significant side effects. The most common are suppression of appetite, interference with sleep, and jittery feelings, mainly due to augmentation of the actions of serotonin and norepinephrine. Some children develop involuntary movements, called tics, which may be due to increased actions of dopamine in circuits that control movement. Although many children are not troubled by these side effects, some cannot tolerate these drugs.

A recent study of about six hundred children[†] (80 percent boys) with ADHD that was sponsored by the National Institute of Mental Health (NIMH) has confirmed the effectiveness of drug treatment not only for days or a few weeks but also for more than a year. Children treated with either methylphenidate or amphetamine (mainly methylphenidate) "showed significantly greater improvement than those given intensive behavioral treatment and community care" for the fourteen months of the study. Like Bradley before them, the researchers who conducted this study were surprised at the results. One of them, Peter Jensen, summed it up this way:[†] "We were surprised how much more effective medication was than even our intensive behavioral treatments in reducing ADHD symptoms. We were also surprised to find that medication combined with behavioral therapy, considered the treatment standard for ADHD, was no more effective than medication alone in reducing ADHD symptoms."

The repeated demonstration that stimulant drugs can have beneficial effects on ADHD has led to extensive testing of schoolchildren for this disorder. At present almost half of referrals to child psychiatrists are for evaluation and treatment of ADHD. In some American school districts one out of five boys is identified as having ADHD and is being treated with amphetamine or methylphenidate. Although many of these children do not have all the features required to make this diagnosis, some psychiatrists are so convinced of the value of these medications that they write prescriptions for what they call "lesser forms of the disorder." Some parents push for the diagnosis because they are convinced that stimulants provide a competitive advantage that they want their children to enjoy.

There is, in fact, evidence that stimulants influence the behavior of all children in the same ways as for those diagnosed with ADHD. A notable example comes from the work of Judith Rapoport and her colleagues at NIMH. They compared the effects of a single oral dose of dextroamphetamine or placebo on normal and hyperactive prepubertal boys, using a double-blind procedure.

Activity was measured with an acceleration-sensitive device that the boys wore on their backs. Measurements made for two hours, starting half an hour after taking the drug or a placebo, showed that drug reduced the activity of the hyperactive and the normal boys by about the same amount. In the case of the hyperactive boys the drug reduced their activity to the level of the unmedicated normal boys, but the activity of the normal boys treated with the drug also declined substantially compared to their activity without it.

Similar results were found in tests of learning and memory. For example, in one test normal boys had an average score of 5.9 after taking a placebo and 8.7 after taking dextroamphetamine; hyperactive boys had a score of 2.9 after taking placebo and 5.9 after taking the drug. Thus the drug improved the performance of the hyperactive boys to the level of the unmedicated normal boys, but it also produced substantial improvement in the performance of the normal boys. As the authors point out: "our data do not strongly support the concept of a unique stimulant drug response[†] in [the ADHD] group."

The evidence that normal children and those diagnosed with ADHD have similar responses to amphetamine does not mean that there is no such thing as ADHD. Everyone who has worked with children recognizes the pattern of behavior that is identified in *DSM-IV*. Everyone also recognizes that the propensity to behave in this way probably reflects a variety of environmental and biological factors, and acknowledges the difficulty of drawing a sharp line between children who belong in this category and their less active and more attentive peers. Some children clearly belong in the ADHD group, the vast majority of children clearly do not, and some are in between. In considering the treatment of these in-between children, however, it is important to remember that the effectiveness of the stimulants has been demonstrated only in studies of children who clearly fit the definition of ADHD.

::

Uncertainty about where to draw the line between normality and pathology, and whom to treat with medications, is not unique to children with ADHD. It has also been extended to a similar pattern of behavior in adults. Largely ignored until the late 1970s, it is now generally accepted that some children with ADHD remain inattentive and impulsive into adulthood.[†] Furthermore, the limited studies of stimulant treatment of adults with ADHD show that the drugs have the same benefits and side effects as those observed with children.

As with other psychiatric disorders that are helped by medications, studies of the effects of amphetamine and methylphenidate on the brain have led to speculations about the biological basis of ADHD that are so tantalizing that they are frequently put forth as fact. For example, in *Driven to Distraction*, a popular book about ADHD, Edward Hallowell and John Ratey conclude that "the medication works by correcting a chemical imbalance of neurotransmitters that exists in ADD in the parts of the brain that regulate attention, impulse control and mood." The idea that stimulants correct a chemical deficiency, especially a deficiency of dopamine, has also captured the public imagination, and is frequently mentioned in newspapers and magazines. Writing in the February 15, 1999, issue of *The New Yorker*, Malcolm Gladwell explains it this way:

Dopamine is the chemical in the brain—the neurotransmitter—that appears to play a major role in things like attention and inhibition. When you tackle a difficult task or pay attention to a complex social situation, you are essentially generating dopamine in parts of the brain that deal with higher cognitive tasks. If you looked at a thousand people at random, you would find a huge variation in their dopamine systems, just as you would if you looked at, say, blood pressure in a random population. A.D.H.D., according to this theory, is the name we give to people whose dopamine falls at the lower end of the scale, the same way we say that people suffer from hypertension if their blood pressure is above a certain point. In order to get normal levels of attention and inhibition, you have to produce normal levels of dopamine.

But dopamine is not the only neurotransmitter that is affected by amphetamine and methylphenidate, and there is evidence that beneficial effects of these drugs are also due to changes in signaling by serotonin and norepinephrine. A role for norepinephrine is supported by the finding that guanfacine (Tenex), a drug that binds to and directly stimulates norepinephrine receptors in the brain, is a useful treatment for ADHD[†] both in children and in adults. So too is desipramine,[†] the drug derived from imipramine, which works primarily by blocking reuptake of norepinephrine and has little effect on dopamine. For example, in a six-week double-blind, placebo-controlled study, that old-fashioned drug decreased hyperactivity, impulsiveness, and inattentiveness in 68 percent of adults with ADHD, whereas none of the subjects had a comparable response to the placebo.

The claims that the medications for ADHD work by correcting a chemical deficiency come from the desire to provide a rationale for their effectiveness. But, as with schizophrenia and depression, there is no direct evidence for an imbalance of a neurotransmitter in people with ADHD, and such claims should not be used to justify the prescription of stimulant drugs. The reason to keep using amphetamine, methylphenidate, guanfacine, or desipramine to treat ADHD is the same as the reason to keep using antipsychotic drugs for schizophrenia or SSRIs for depression—because controlled clinical studies show that in certain cases they have some beneficial effects.

PART II :: BETTER PSYCHOTHERAPEUTIC DRUGS

It has been both our blessing and our curse that we had effective drug therapy for emotional disorders before we had a science of behavioral pathology. Our best hope for getting better psychotherapeutic drugs is to understand better the causes of emotional disorders.

—Leo Hollister (1975)

6 :: *Martha's Panic*

A lady, age 20, became affected with some symptoms which were sup-
posed to be hysterical. . . . After she had been in this nervous state about
three months it was observed that her pulse had become singularly rapid
. . . never under 120 and often much higher . . . [and] her eyes assumed a
singular appearance for the eyeballs were apparently enlarged, so that
when she slept or tried to shut her eyes they were incapable of closing.
—Robert Graves (1835)

Twenty-five years after Paul Charpentier's creation of chlorpromazine, the *Journal of the American Medical Association* asked Leo
Hollister, a physician at the Veterans Administration Hospital in Palo
Alto, California, to write a review article called "Drugs for emotional
disorders." Hollister was chosen for this prestigious assignment
because he had done pioneering work on the clinical assessment of
these medications. The article he prepared explained how chlorpromazine, imipramine, diazepam, and amphetamine had improved
the practice of psychiatry.

But Hollister also informed the readers of this widely read journal about the imperfections of these drugs and the need to make
substantial improvements. His goal was what he called an "ideal psychotherapeutic drug,"[†] one that corrects the brain abnormalities that
cause mental symptoms, instead of just covering them up—a goal
that would be achieved only when these causes are identified. As

Hollister put it at the end of his article: "Our best hope for getting better psychotherapeutic drugs is to understand better the causes of emotional disorders."

Hollister's last sentence rings as true today as it did when he wrote it. But the causes of the disorders that he was mainly interested in, such as depression and schizophrenia, have been very difficult to find, and appear to be more complicated than he imagined. Discovering them depends on new investigative techniques that are only now becoming available.

There are, however, a few psychiatric disorders whose origins are already known and whose pharmacological treatment is guided by this information—just the sort of thing that Hollister had in mind. In this chapter I will describe one of these disorders and the discoveries of some of the steps that give rise to it, which Hollister would group together as "causes." By providing details of this successful process of discovery, including some of its surprises and blind alleys, I will illustrate the way that research of this type progresses, and the way that it influences the treatment of patients. But first I will show you the human face of this disorder by telling you the story of Martha's panic.

::

I first learned that Martha was in a state of panic from her husband, Trevor, a family friend, who was so disturbed by her distress that he called me from Hawaii, where they were vacationing. Trevor was particularly alarmed because Martha kept staring at him with a fearful expression and because her heart was beating so fiercely that he could see its banging through her clothes. The hotel doctor was also alarmed, and advised him to take her home immediately for psychiatric treatment. Trevor asked if I would help arrange for her care.

Trevor and Martha had met as students at a large eastern university. She, a twenty-six-year-old petite blonde from an old Kentucky family, had been to college in the South and was well on her way to an MBA. He, big and brawny, three years older, grandson of Irish immigrants, was in law school. They married when Trevor graduated, then moved to California, where he had been offered a job by a small law firm.

Both thrived in this new environment. Trevor, at thirty-eight, was becoming established in the firm that had recruited him. Martha had found a position as a financial officer in a tiny biotechnology company. She was feeling sufficiently settled to decide to have a baby.

It was in the wake of this decision that Martha's nervousness first appeared. Normally steady and unflappable, she suddenly found herself easily offended, edgy, and prone to tears. Sleep was erratic: she kept waking up in a sweat. There was also an uncomfortable sense of fullness in her neck, as if she were choking. Noticing that Martha seemed very tired when she came to work, her boss suggested she take a vacation. They chose Kauai.

But instead of getting better on this holiday, Martha got worse. Luxuriously settled in a room overlooking the Pacific, she would toss and turn all night, her mind racing. Having gone to Kauai in search of romance, Martha found that she no longer had any sexual interest in her husband. As she tried to explain this to Trevor, Martha became terrified—heart thumping, hands shaking, drenched in sweat. For the first time in her life this confident and optimistic woman felt helpless and out of control. Two days later she was in my office.

::

The extensive knowledge we now have about Martha's disorder can be traced back to the work of a nineteenth-century British physician, Caleb Parry. Although he was a busy society doctor in Bath, then a fashionable spa, Parry also found the time to write articles about his medical observations, some of which his son published after his death in 1825. Among these was a paper about patients with pounding hearts, swollen necks, and prominent staring eyes.

Noticing such combinations of symptoms and physical abnormalities—combinations called syndromes—was, and remains, a crucial first step in the identification and understanding of specific diseases. Ten years after Parry's posthumous publication, the same combination was observed at the Meath Hospital in Dublin by Robert Graves, a legendary Irish physician of his time. In 1840 Carl Basedow, a German doctor, also called attention to an association between heart palpitations, a swelling in the neck, and prominent staring eyes. Together these observations were the basis for the naming of a new disease—Parry-Graves-Basedow disease, now generally known as Graves' disease.

But having a name for the disease does not explain its causes. In medicine these causes are thought of in terms of a series of biological processes that lead from health to an abnormal state of being. These processes may be roughly subdivided into those that initiate a disease, called pathogenesis, and those that give rise to symptoms, called pathophysiology—although the line between these processes is rarely clear. The important point is that a disease can be thought of *as a series of biological processes* and that any of them can become a target for a treatment, because interfering with just one of them may relieve symptoms.

In the case of Graves' disease the main contribution that Parry, Graves, and Basedow made to an understanding of its pathophysiology is that each of them recognized that it somehow affects the thyroid, the H-shaped gland that sits immediately below the Adam's apple. But none of them understood how this was related to their patients' pounding hearts. To Graves it seemed likely that it was the pounding heart that made the thyroid swell, by engorging it with blood. Were this true, the main target for treatment would be the heart rather than the thyroid.

It was not until the end of the nineteenth century that Graves' proposal was turned on its head, and the pounding heart was shown to be the consequence—instead of the cause—of the swollen thyroid. This reversal was based on experiments showing that the thyroid of patients with Graves' disease secretes too much of a hormone called thyroxine and that the excess thyroxine makes the heart beat faster and more vigorously. But the reason that the patients were secreting too much thyroxine remained a mystery. So too did the reason for their staring eyes.

Nevertheless, as with many other illnesses, this lack of a complete understanding did not preclude the development of effective treatments. Once it became apparent that a patient with Graves' disease was being poisoned by excessive secretion of thyroxine from the thyroid, surgeons began removing large portions of the enlarged gland—so accessible to their scalpels—with the aim of reducing the amount of thyroxine to normal levels. Although the surgeons frequently removed too much of the thyroid, bringing the patient from a condition of hormone excess (hyperthyroidism) to a condition of hormone deficiency (hypothyroidism), a remedy for this overzealous surgery was available. The deficiency (which also gives rise to mental symptoms—in this case apathy and depression) could be corrected by taking thyroxine. This was easily achieved with daily doses of thyroxine-containing pills made from extracts of animal thyroids, a by-product of the meatpacking industry.

In addition to hypothyroidism due to removal of too much thyroid tissue, surgery has other drawbacks. Among them is an unsightly scar of the neck, occasional damage of nerves that control the vocal cords, and the ever-present danger of death on the operating table. For those wishing to avoid surgery, an alternative was introduced in the 1940s: drugs that block the manufacture of thyroxine[†] by the thyroid gland.

Like many other drugs, such as ephedrine and reserpine, the discovery of drugs that block the manufacture of thyroxine can be traced to an observation about the actions of a plant—in this case cabbage. Scientists happened to notice that laboratory rabbits that were fed mostly cabbage had very large thyroids. But in contrast with the large thyroid of Graves' disease, which makes enormous amounts of thyroxine, the large thyroid of a cabbage-fed rabbit made very little thyroxine.

The cause of the low levels of thyroxine in the cabbage-fed rabbits was soon traced to chemicals in the vegetable that block an enzyme, called thyroid peroxidase, that plays a critical role in the manufacture of the hormone. Identification of the chemicals in the cabbage eventually led to the creation of propylthiouracil and methimazole, two valuable drugs that are still taken every day by vast numbers of people as a treatment for Graves' disease. Some lucky patients can give up their pills eventually, with no return of symptoms. Most must continue to take them or decide instead to have a large fraction of their thyroid gland physically destroyed either surgically or by radiation.

::

Although I had been well aware of Graves' disease for many years, it was not what had popped into my mind when I first heard Trevor's description of Martha. Instead I assumed she had a condition called panic disorder, which is much more commonly seen in a psychiatric practice. Patients with this disorder have episodes of intense fear, called panic attacks, as well as sustained periods of anxiety. Like Graves' disease, panic disorder often strikes young women in their thirties. Martha's most distressing symptoms—such as her thumping heart, her sensation of choking, her extreme fear, and her sense of losing control—are characteristic features of panic disorder.

What pointed me in the right direction were Martha's staring eyes. Trevor had mentioned them in passing when he called from Hawaii. But it was only when I met Martha face-to-face that I realized that they were actually protruding slightly from their sockets— a telltale sign. One glance at those eyes and I was transported back to the Thyroid Clinic at the National Institutes of Health, where I had worked in the early 1960s. Martha's protruding eyes were like those of so many other young women who were referred to the Thyroid Clinic for diagnosis and treatment. All of them had Graves' disease.

So did Martha. In addition to her characteristic stare, the swollen thyroid in her neck was clearly visible. Along with her thumping heart and the symptoms that Trevor had already reported, the diagnosis seemed clear. Confirmation would come from a direct measurement of the thyroxine in her blood by way of a simple blood test.

But even before this confirmation, I felt sufficiently confident about the diagnosis to give Martha a drug called propanolol, to help relieve her feeling of panic. Propanolol is a member of a class of drugs known as beta-blockers, which interfere with the action of adrenaline on receptors called beta-receptors. It helps people with Graves' disease because many of their symptoms, such as rapid heart rate, are due to thyroxine-induced amplification of the effects of the small amount of adrenaline that is always present in the blood. This amplification mimics the response to the large amount of adrenaline that pours out of the adrenal gland in a frightening situation, and propanolol can block it.

In Martha's case the propanolol started working in just a few minutes. One pill—along with my explanation that she had a well-understood and readily treatable condition—brought substantial relief. To help calm her down I also gave Martha some Valium. The combined effects of these drugs reduced her distress while she awaited further treatment.

That treatment was provided by the thyroid specialist to whom I referred her. Once he received the results of a blood test that showed that Martha's thyroid gland was indeed making too much thyroxine, he prescribed methimazole, which slows down the manufacture of this hormone. Over the next few weeks he adjusted the dose of this drug on the basis of more blood tests, and succeeded in bringing her level of thyroxine back to normal. He also stopped the propanolol and the Valium, both of which became unnecessary when her thyroxine levels fell. Within a month Martha had regained some of the weight she had lost due to hyperthyroidism, and was feeling like her old self.

::

The discoveries about the role of thyroxine in Graves' disease illustrates two practical benefits of understanding the pathophysiology of this disorder. First, the diagnosis can be confirmed by measurements of thyroxine in the blood and need not rely solely on a description of symptoms and a physical examination. Second, the treatment can be based on knowledge of the biological processes that give rise to symptoms—in this case giving methimazole to block excessive manufacture of thyroxine. Because methimazole works directly on thyroid peroxidase, an enzyme that is needed to make thyroxine, the dose can be adjusted in a straightforward manner, based on sampling of the blood levels of the hormone.

But the discovery of the role of thyroxine in Graves' disease does not explain how the disease process gets started. Why did Martha's thyroid, which made normal amounts of thyroxine for the first thirty-five years of her life, suddenly go into high gear?

In the middle of the twentieth century it was proposed that the cause of the excessive thyroxine production could ultimately be attributed to sustained emotional distress.[†] This idea was popularized in 1950 by Franz Alexander, the director of the Chicago Institute for Psychoanalysis, in his influential book, *Psychosomatic Medicine*. It was based on discoveries about the normal regulation of blood thyroxine levels by the brain and the pituitary gland, and by feedback to the brain and pituitary. Put succinctly:

1. Thyroxine production is controlled by thyroid-stimulating hormone (TSH), which binds to a receptor on the thyroid gland.
2. TSH comes from the pituitary gland, the gland that sits at the base of the brain.
3. Release of TSH from the pituitary is, in turn, controlled by another hormone, called TRH, from a part of the brain called the hypothalamus.
4. The level of thyroxine in the blood regulates the manufacture of TSH and TRH by feedback to the brain and pituitary.

Once it became clear that the brain plays an important role in the control of the thyroid gland, it seemed plausible that this chain of command could be controlled by emotions. In this scenario sustained emotional distress might raise the level of TRH, signaling the pituitary to pump out a lot of TSH. If the emotional distress maintained this pumping of TSH despite feedback signals from the rising blood levels of thyroxine, the result would be the hyperactive thyroid of Graves' disease.

What made this proposal so interesting is that it could be evaluated by measuring the blood levels of TSH in patients with Graves' disease. If it is correct, their blood should contain an abnormally large amount of TSH. To test this proposal, D. D. Adams and H. D. Purves, of the University of Otago in New Zealand, developed a method for measuring TSH in human blood by injecting blood serum into a guinea pig and measuring its stimulant effect on the secretion of thyroxine by the guinea pig's thyroid gland. In 1956 they reported that serum from Mrs. McCabe[†], a patient with Graves' disease, was a much more potent thyroid stimulant than normal human serum. This is exactly what would be expected if Graves' disease is the result of a hyperactive brain and pituitary. It looked as if Franz Alexander was right.

But one detail didn't fit. Although Mrs. McCabe's serum did indeed stimulate the thyroid gland of the guinea pig, its effect lasted much longer than that of authentic TSH that had been extracted from pituitary glands. This suggested that the relevant material in Mrs. McCabe's serum might be a novel substance that can also stimulate the thyroid. Because the distinguishing feature of this new substance was its prolonged effect, they called it long-acting thyroid stimulator (LATS).

The identity of LATS, and the way that it affects the thyroid,[†] were worked out in the ensuing decade by Adams, Purves, and other researchers. To their great surprise they found that LATS belonged to a class of proteins called antibodies, which includes the protective proteins we make to ward off infection by foreign invaders, such as bacteria. The odd thing about this antibody is that, instead of binding to a foreign substance, it binds to a normal component of the body—the receptor protein on thyroid cells that normally binds TSH and controls the manufacture of thyroxine. By binding this receptor, the antibody (i.e., LATS) tricks the thyroid cell into responding as if it is being stimulated by TSH.[†]

Because Graves' disease is the result of the manufacture of antibodies that attack the thyroid, it is now classified in a category called autoimmune diseases, which also includes rheumatoid arthritis (in which antibodies attack the linings of joints) and multiple sclerosis (in which antibodies attack the tissues around nerve fibers). In each of these diseases the immune system makes antibodies that attack a person's own proteins instead of the proteins of foreign invaders.

The specific proteins that are the targets of these antibodies, and the tissues in which they are displayed, determine the resultant symptoms.

The finding that Graves' disease is an autoimmune disease also helped explain another of its characteristic features—the staring eyes—because it led to the discovery that these patients also make antibodies that bind to cells in the eye sockets.[†] These antibodies produce swelling of the tissues in the eye sockets, which makes the eyes bulge. This explains why these eye problems, which are an independent manifestation of autoimmunity, may persist after the excessive production of thyroxine has been brought under control with drugs. Understanding the pathophysiology of the protruding eyes raised the possibility of developing a treatment that blocks the manufacture or actions of the relevant antibodies, but that goal has not yet been achieved.

Nor has the goal of fully understanding the pathogenesis of Graves' disease. Nevertheless, the discovery of the role of autoimmunity in Graves' disease does provide a new way of thinking about it. It tells us that prevention of the disease won't come from psychotherapy, as Alexander believed. Instead it will come from finding a way to stop the formation of the pathogenic antibodies.

Yet answering those questions just shifts the mystery back one step. Why do people with Graves' disease make antibodies to their own body proteins, while the rest of us don't? Why do they make antibodies that specifically attack the thyroid and the tissues around the eyes, instead of the joints (as in rheumatoid arthritis) or the tissues surrounding nerve fibers (as in multiple sclerosis)?

::

One answer that many have proposed is heredity. Franz Alexander, for example, was well aware that there were many people who display the psychological features that he considered characteristic of Graves' disease, yet never develop hyperthyroidism. His explanation, in 1950: "Heredity, of course,[†] may be a decisive factor. Most authors recognize, besides environmental influences, a hereditary factor in an inherited susceptibility to hyperthyroidism."

In support of the claim that susceptibility to Graves' disease is inherited is the evidence that it runs in families. But the clustering of Graves' disease in certain families is far less dramatic than the familial clustering of diseases that are caused by inheritance of a single abnormal gene. In some hereditary diseases of this kind, half (on average) of the children (both male and female) of an affected parent will develop the same disease, because each child has a fifty-fifty chance of inheriting the abnormal gene. Furthermore, half of the siblings of an affected person (both male and female) will also have the disease, because each sibling has a fifty-fifty chance of sharing the abnormal gene. Among the diseases that show this pattern of inheritance are rare forms of Alzheimer's disease, which I will turn to later.

The pattern of hereditary transmission of susceptibility to Graves' disease is much less straightforward. One complicating factor is gender. Whereas about 2 percent of women develop Graves' disease, only one-tenth as many men do. For this reason studies of familial clustering of this disease tend to focus on women. In the largest study of this kind, which included 322 Hungarian women with Graves' disease,[†] 6 percent of their daughters and 7 percent of their sisters were shown to also have this disease. Although the affliction of 6–7 percent of the close relatives of people with Graves' disease is considerably greater than the prevalence of this disease in the overall population, supporting the importance of heredity, it is a far cry from a prevalence of 50 percent. This indicates that Graves' disease is not the simple consequence of inheritance of a single abnormal gene.

Another reason to consider other causes is that clustering in families could reflect shared environment as well as shared genes.[†] For example, members of the same family tend to infect each other with contagious infectious diseases, including those that attack the thyroid gland. There is even some evidence that bacterial or viral infections of the thyroid gland may trick the immune system into making antibodies against the gland itself. If this is the case, it might lead to Graves' disease.

Nevertheless, some geneticists are so persuaded of the importance of genes[†] that they are hard at work on this problem. The reason they are undeterred by the fairly low level of familial clustering found in Graves' disease is that this pattern is actually quite characteristic of many other human diseases, such as diabetes and high blood pressure, for which some genetic basis has already been established. Such diseases—which geneticists call complex genetic diseases—reflect interactions between multiple genetic and environmental factors. This category also includes mental disorders such as schizophrenia and bipolar disorder, which show a familial clustering that is greater than that of Graves' disease. Even milder psychiatric disorders, such as Clara's, have some genetic basis, which I will come to later. Identification of the relevant genes will provide clues to the pathophysiology of these complex diseases, which in turn will lead to new treatments.

::

Very few people who consult psychiatrists have disorders that are as well understood as Martha's. It can therefore be argued that Martha's case is fundamentally different from Clara's. Shouldn't Martha's case be viewed as a medical disorder or a physical disorder rather than a psychiatric disorder or a mental disorder?

For much of the twentieth century leading psychiatrists did indeed make this distinction between psychiatric disorders and medical disorders.[†] Believing that mental disorders are caused by traumatic life events and can be effectively treated only by psychological means, many psychiatrists were content to separate themselves from the rest of medicine. In their view, Martha's case did not really belong in the domain of psychiatry.

But this distinction has withered away. The clearest evidence of this comes from *DSM-IV*, the latest edition of the American Psychiatric Association's handbook of diagnoses. Published in 1994, it advocates a broad definition of mental disorders and a blurring of their distinction from physical disorders:

> Although this volume is titled the *Diagnostic and Statistical Manual of Mental Disorders*, the term *mental disorder* unfortunately implies a distinction between "mental disorders" and

"physical disorders" that is a reductionistic anachronism of mind/body dualism. A compelling literature documents that there is much "physical" in "mental" disorders and much "mental" in "physical" disorders. The problem raised by the term "mental" disorders has been much clearer than its solution, and, unfortunately, the term persists in the title of *DSM-IV* because we have not found an appropriate substitute. Moreover, although this manual provides a classification of mental disorders, it must be admitted that no definition adequately specifies precise boundaries for the concept of "mental disorder." The concept of mental disorder, like many other concepts in medicine and science, lacks a consistent operational definition that covers all situations.

Based on this reasoning, *DSM-IV* encompasses all disorders with prominent mental symptoms, including those also claimed by other branches of medicine. It explicitly includes cases such as Martha's, giving her the official psychiatric diagnosis of "anxiety disorder due to thyrotoxicosis, with panic attacks"—*DSM-IV* diagnostic code number 293.89.

The important feature that distinguishes Martha's diagnosis from most others in this manual is the words "due to." In contrast, most other diagnoses in *DSM-IV*, such as dysthymic disorder and body dysmorphic disorder (Clara's diagnoses), are not due to known factors. Instead they resemble the diagnoses that were developed in the days of Parry and Graves and Basedow—diagnoses based exclusively on patterns of symptoms that have been tentatively lumped together as syndromes.

The mental disorders defined by some of these diagnoses also resemble Graves' disease in other ways. As with Graves' disease, there is evidence that combinations of gene variants and environmental factors play a part in their development. As with Graves' disease, there are useful drug treatments. And, as with Graves' disease, these drug treatments don't always work. In the case of Graves' disease a measure of the shortcomings of drug treatment comes from a recent study by a group of Robert Graves' Irish countrymen.[†] Published in 1999, their ten-year follow-up of drug treatment for Graves' disease at Cork University Hospital showed that only about a third of their patients had a sustained remission of thyroid hyperactivity. The other two-thirds had recurrent hyperthyroidism that often necessitated physical destruction of the thyroid by surgery or radiation.

The other big problem in treating Graves' disease—whether with drugs, surgery, or radiation—is that none of the treatments interferes with the persistent production of pathogenic antibodies. In some cases the antibodies that attack the thyroid eventually lead to the destruction of much of the gland, resulting in inadequate thyroid function (hypothyroidism) instead of excessive function (hyperthyroidism). Fortunately hypothyroidism can be treated with thyroxine. In other cases antibodies that attack the cells in the eye sockets can lead to such extensive bulging that eye functions can be impaired. X-ray treatments of the eye sockets[†] have been used to reduce the swelling, but the results are frequently disappointing.

Luckily Martha has avoided such complications. After three years of treatment with methimazole her blood levels of thyroxine continue to be normal, and there has been no recurrence of her behavioral disturbances. Furthermore, her eyes seem a bit less prominent than they did when she first came to my office, and function normally.

Nevertheless, Martha shares Clara's longing for better treatments. Both are aware that the pills and other remedies that are available to them have limitations. Although the treatment of Graves' disease is based on a much better understanding of pathophysiology than the treatment of dysthymia and body dysmorphic disorder, there is still a great deal of room for improvement.

For both Martha and Clara the next therapeutic advances may be based on a new approach—the identification of genes that play a role in the development of their disorders. Until recently there was no practical way to accomplish this daunting task. But techniques have recently been developed to find the gene variants that make some people susceptible to such complex abnormalities. In the next chapter I will explain not only how this is being done, but also how genetic discoveries can become the basis for novel medications.

Comparison of Martha's and Clara's Disorders

	Martha's	Clara's
Causes	Unidentified genes and environment	Unidentified genes and environment
Pathophysiology	Autoantibodies; increased manufacture of thyroxine; increased sensitivity to adrenaline	Unknown
Drugs	Methimazole (blocks manufacture of thyroxine) Propanolol (blocks action of adrenaline)	SSRIs and others (not clear how they work)
Results of Treatment	Variable	Variable
Next Big Steps	Clarify pathophysiology Find genes	Clarify pathophysiology Find genes

7 :: Gene Variants

At one end of the series[†] [of people with mental disorders] stand those extreme cases of whom one can say: These people would have fallen ill whatever happened, whatever they experienced, however merciful life has been to them.

—Sigmund Freud (1917)

In the autumn of 1901 a senior physician at the Hospital for the Mentally Ill and Epileptics in Frankfurt, Germany, examined a middle-aged woman named Auguste D.,[†] whose main symptoms were depression, confusion, paranoia, and a progressive loss of memory. So intrigued was he with the strange behavior of this patient that he stayed in touch with her even after he had taken a new position at the Royal Psychiatric Clinic in Munich. When she died in 1906, at the age of fifty-one, he had her brain sent to him for examination.

A specialist in the microscopic study of the nervous system, he was perplexed by what he found. Many of the nerve cells in Auguste D.'s brain had degenerated and left behind masses of tangled fibrils. Furthermore, scattered throughout the brain, including its outer layer, called the cortex, were tiny clumps of a substance now called amyloid, whose origin he could not explain. Half a year after the death of Auguste D. he presented his findings at the 37th Conference of South West German Psychiatrists in Tübingen. He published them in 1907 in *Allgemeine Zeitschrift für Psychiatrie*, a German psychiatric journal, entitling his article "A new disease of the cortex."

Within a few years the same abnormalities were observed in the brains of several more people who had suffered from severe memory impairment and other symptoms like Auguste D.'s. In 1910 Emil Kraepelin, director of Munich's Royal Psychiatric Clinic and author of *Handbook of Psychiatry,* the leading diagnostic manual of its time, summarized these findings in the newly published eighth edition. In honor of his colleague, Alois Alzheimer, who had discovered this new form of brain disease, Kraepelin named it Alzheimer's disease.

::

The discovery of amyloid in the brains of patients with Alzheimer's disease was an important step in our present understanding of this disorder. But little progress was made for more than half a century after Alzheimer's initial report. Although there was speculation that the accumulation of amyloid gives rise to the characteristic mental symptoms—just as the accumulation of thyroxine in Graves' disease gives rise to its set of characteristic mental symptoms—no one knew how to test this idea. Unlike thyroxine, which can be injected into experimental animals to simulate the symptoms of Graves' disease, at the time there was no way to introduce amyloid into the brains of experimental animals to find out what happens.

The main reason amyloid is so much harder to work with than thyroxine is that it is such a complex biological material. Whereas thyroxine is a small molecule that is easily purified from the thyroid gland, amyloid is a mixture of proteins that are hard to purify from the brain. Furthermore, the chemical structure of thyroxine could be readily determined with the techniques of the early twentieth century, whereas deciphering the elaborate chemical structures of amyloid's components depended on the development of an entirely new technology.

As in the studies of the structure of thyroxine, there were two main phases in the studies of the structure of amyloid. The first was to break it down into components that are small enough to analyze. The second was to find out how the components are put together to make the final unique product. A breakthrough came in the early 1980s[†] when a fragment of a protein was purified from amyloid and its building blocks were identified. There are twenty different amino acid building blocks that are assembled to make all proteins, each symbolized by a single letter, such as *V* for valine and *I* for isoleucine. The protein fragment found in Alzheimer's disease—now called A-beta—contains a particular set of amino acids.

But knowing which amino acids A-beta contains tells little about the biological information it conveys. Just as the letters *a*, *p*, and *t* can be strung together in different orders to produce words as different as *pat* and *tap,* so too can the twenty amino acid building blocks used to make proteins be strung together in different orders—like letters in words—to make proteins with different functions. In the case of A-beta, application of new techniques for sequencing fragments of protein soon revealed that it is a "word" that is forty letters long, with the amino acid at each position contributing to the overall properties of the fragment.

Once the arrangement of the amino acid "letters" in A-beta was found, attention turned to the origin of this major component of amyloid. We now know that A-beta is a fragment of a much larger brain protein, and that this fragment, and a slightly longer version, are made by cutting up this protein with specialized brain enzymes that work as molecular scissors. The details of the structure of the complete protein were discovered with the aid of another major chemical technology that was developed in the latter part of the twentieth century.

That new technology is based on the fact that the order of the amino acid building blocks in each protein is determined by the order of the building blocks in the DNA of the particular gene that encodes this protein. DNA itself is made up of four simple building blocks, called nucleotides, symbolized by the letters A, C, G, and T; a string of three consecutive nucleotides in DNA is the code for a particular amino acid. For example, the DNA code for the amino acid valine (V) is CAA,[†] and the DNA code for amino acid isoleucine (I) is TAA. Knowing the sequence of amino acids in A-beta, researchers found it was possible to search through human

DNA to find the long string of nucleotides that encodes it. This string of nucleotides is contained within the gene that encodes the entire protein—called amyloid precursor protein (APP)—from which A-beta is derived. Figuring out the sequence of As, Cs, Ts, and Gs in this gene, and knowing the DNA code for each amino acid, revealed the sequence of the amino acids in the complete APP. Thus, by combining these techniques, the structure of the gene that encodes APP, the structure of the protein itself, and the structure of the protein fragment in amyloid became known—setting the stage for further research on the origins of Alzheimer's disease.

::

While these structures were being deciphered with the aid of these chemical techniques, another clue to the origin of Alzheimer's disease was found: Alzheimer's disease is extremely common in the members of certain families. This is especially true of rare cases like Auguste D.'s, in which symptoms appear at a relatively young age, often in the forties. Siblings and children of a person with this form of the disease—called early-onset Alzheimer's disease—have about a fifty-fifty chance of also being affected. In contrast, the common variety of Alzheimer's disease—called late-onset Alzheimer's disease because it tends to appear after age sixty and becomes increasingly common with advancing age—is much less concentrated in particular families.

Knowing that early-onset Alzheimer's disease affects about half of the children and siblings of an affected person tells us a lot about its hereditary basis. To understand how geneticists think about this, it is necessary to review a few facts about human genetics.

1. There are about thirty thousand human genes,[†] each of which consists of a certain sequence of the four basic building blocks of DNA.

2. There are alternative forms of each gene called alleles, or gene variants. The variations that define these alternative forms may be as small as a difference in a single building block in the DNA sequence or as large as the complete absence of the entire gene.

Such variations may influence the structure and function of the protein that the gene encodes, and the amount that is made—even whether the protein is made at all.

3. Gene variants may manifest themselves as differences in human physical and behavioral traits. Some gene variants contribute to susceptibility to a disease, or, rarely, may even cause a disease.

4. A person inherits two copies of each gene, one from each parent.

5. Because a parent may have two different variants of a particular gene—say, variant A and variant B—each child has a fifty-fifty chance of inheriting variant A and a fifty-fifty chance of inheriting variant B from that parent. If one of these variants—say, variant A—causes a disease, those children who inherit variant A will get this disease, whereas those who inherit variant B will be spared.

From these facts it should be clear why geneticists take notice when they learn about a disorder, such as early-onset Alzheimer's disease, that is passed down by an affected person to about half of his or her children. The inference they make is that the disorder is caused by a gene variant that transmits the disease to those unlucky children who happen to inherit that particular variant. This clear-cut form of inheritance is called Mendelian inheritance in honor of Gregor Mendel, the Czech priest who discovered the basic principles of heredity in the nineteenth century. Knowing the DNA structure of the gene for APP, scientists began hunting for variations in its structure in the DNA of people with early-onset Alzheimer's disease.

Success came in 1991, when Alison Goate, John Hardy, and their colleagues at St. Mary's Hospital Medical School in London identified a variation in the gene for APP in the DNA from a British patient[†] who had died of familial Alzheimer's disease. Furthermore, the five other affected family members had exactly the same variation. Now named the "London variant," its only abnormality is a substitution of a C for an A at a precise location in the gene. This causes a substitution of the amino acid "letter" valine for the amino acid "letter" isoleucine at amino acid position 717 in the string of amino acids that make up APP—a position that is close to one end of A-beta. The discovery that this substitution is found only in people who eventually develop early-onset Alzheimer's disease suggests that the substitution leads to an increased accumulation of A-beta, and that the accumulation of this fragment of APP somehow leads to the dysfunction and degeneration of nerve cells that are the basis of the symptoms of Alzheimer's disease.

::

Confirmation of this discovery followed quickly. Over the next few years the same substitution in the APP gene was found in people with early-onset Alzheimer's disease from more than a dozen different families in Europe and Japan. Furthermore, some other families had different changes in the gene for APP,[†] including one at the other end of A-beta, called the "Swedish variant." These additional findings supported the proposal that subtle alterations in the structure of APP are one way to get an accumulation of the amyloid that Alzheimer had observed.

But it soon became apparent that only about one out of ten families with early-onset Alzheimer's disease have variants of the APP gene, indicating that variants of other genes must also be responsible for exactly the same disorder. This was confirmed in 1995 with the discovery of a variation in a different gene that causes Alzheimer's disease in members of another family. In this case the gene variant was discovered by comparing the DNA of those family members who developed the disease with the DNA of their unaffected relatives. The gene itself had not even been noticed until then.

Variants of this same gene were soon found in about half of the families with early-onset Alzheimer's disease, leading to the naming of this gene presenilin-1 (for a "pre-senile," or early-onset, form of the disease). In the same year a variant of another gene, named presenilin-2,[†] was found in the affected members of other families with familial Alzheimer's disease. As with presenilin-1, there was at the time no inkling of the function of presenilin-2 or the way the variation in its structure causes Alzheimer's disease.

Fortunately, the discoveries that had already been made about amyloid and APP helped scientists figure out the roles of the two presenilins and their pathogenic variants. We now know that both presenilin-1 and presenilin-2 are components of a complex enzyme, called gamma-secretase, that cuts APP, releasing one end of A-beta. The variants of the genes for either of the presenilins change the function of gamma-secretase in a way that somehow leads to increased accumulation of A-beta. It, in turn, is deposited in clumps of amyloid and contributes to the destruction of nerve cells.

In the course of these studies of gamma-secretase and its role in cleaving APP to make A-beta, several other enzymes were discovered that also cut APP[†] in specific ways. One, called beta-secretase, cuts APP at the other end of A-beta, making it gamma secretase's partner in the creation of this peptide. Another, called alpha-secretase, prevents the accumulation of this dangerous material.

::

Which finally leads us to drugs. Knowing about the importance of A-beta has stimulated pharmaceutical companies to search for drugs that prevent its formation. Such drugs would be fundamentally different from those currently used to treat Alzheimer's disease, which do not stop the progression of the disease. Instead, the current drugs just counteract some of the mental symptoms by influencing the activity of a neurotransmitter—using the same approach that is presently being used to counteract the symptoms of other behavioral disorders, such as depression.

In the case of Alzheimer's disease the neurotransmitter that is targeted is acetylcholine,[†] a participant in many brain circuits. It was selected for two main reasons. First, it had been noticed that drugs that block receptors for acetylcholine aggravate the memory impairment of people with Alzheimer's disease. Second, postmortem examination of the brains of people with Alzheimer's disease showed that the degree of memory impairment is correlated with the extent of the degeneration of the nerve cells that make acetylcholine.

To counteract the deficit of acetylcholine, drugs were developed to prolong the activity of the remaining neurotransmitter. The drugs, such as donepezil (Aricept), rivastigmine (Exelon), and galantamine (Reminyl), do their job by blocking the actions of an enzyme called acetylcholinesterase, which degrades acetylcholine, just as the MAO inhibitors used to treat depression block the action of the enzyme that degrades norepinephrine and serotonin. Placebo-controlled studies have demonstrated that treatment with these drugs produces transient improvements[†] in memory and in other brain functions. But as amyloid continues to accumulate and more nerve cells die, the drugs become less effective.

To stop the progression of the disease, pharmaceutical companies are trying to make drugs that prevent the ongoing accumulation of amyloid.[†] Although there is presently no proof that this approach will be successful, they are investing billions of dollars to design such drugs and to test them in patients. They are willing to make this investment because they believe that the new drugs will interfere with a critical step in the pathogenesis of Alzheimer's disease and prevent further mental deterioration.

The main approach that is currently being taken is to create drugs that block the actions of the secretase enzymes that cut up APP to make A-beta. Several pharmaceutical companies have succeeded in making drugs that inhibit beta-secretase, and are studying their effects on patients with Alzheimer's disease. Inhibitors of gamma-secretase have also been invented and are being tested. In addition, other approaches are being used to try to break up the clumps of A-beta and amyloid that have already accumulated. Such drugs might be useful not only for patients with early-onset Alzheimer's disease who have known genetic abnormalities that lead to the accumulation of A-beta and amyloid, but also for those with the much more prevalent late-onset form of the disease.

::

While drugs that inhibit secretases are being created and tested, work continues on the pathogenesis of the late-onset forms of Alzheimer's disease. As with the rare early-onset forms, genes play a part. But unlike the gene variants that cause early-onset Alzheimer's disease, usually by the age of fifty, the gene variants implicated in the late-onset forms only increase the risk of developing the disease in later life; they don't necessarily cause it.

Of the genes that influence this susceptibility to late-onset Alzheimer's disease, the most important, called APOE,[†] encodes apolipoprotein E, a protein that binds to cholesterol. This gene exists in three variant forms called APOE-2, APOE-3, and APOE-4. Unlike the rare gene variants that cause early-onset Alzheimer's disease, the variant form of APOE that increases the risk of the late-onset form is very common.

The initial evidence that implicated the APOE gene in Alzheimer's disease came from a study of DNA samples from elderly people which showed that those with the disease are more likely to have the APOE-4 gene variant. The risk is especially great for those who inherit two copies of the APOE-4 gene variant—one from each parent. They have about a fifty-fifty chance of developing Alzheimer's disease by age seventy. Inheriting one copy of the APOE-4 gene variant (from only one parent) is not as dangerous.

Despite the clear effect of the APOE-4 gene variant on the risk of developing Alzheimer's disease, some people with two copies of APOE-4 don't get Alzheimer's disease even if they live into their nineties. Because the APOE-4 gene variant just increases susceptibility to Alzheimer's disease, but does not cause the disease, it is called a susceptibility gene variant—in contrast with a causal gene variant such as the London variant of the APP gene. Whether or not a person who has an APOE-4 variant develops Alzheimer's disease depends, in part, on the possession of other susceptibility gene variants, several of which are being actively studied. It also depends, in part, on a variety of environmental factors, such as having received repeated blows to the head.

How does inheritance of APOE-4 gene variants increase the risk of having Alzheimer's disease? Clues come from the discoveries that amyloid contains cholesterol and that the APOE-4 protein deposits cholesterol in amyloid, which increases the accumulation of this pathogenic substance. A role for cholesterol in the accumulation of amyloid fits with the observation that people who take a type of cholesterol-lowering drug called a statin, such as simvastatin (Zocor), not only reduce their risk of heart attack but also have less Alzheimer's disease.[†] Although the reason statins reduce the risk of Alzheimer's disease is not yet established, some physicians now prescribe them for patients who have this disease as an off-label treatment that may slow the destruction of brain cells. They may also prescribe ibuprofen (Advil, Motrin), an anti-inflammatory and analgesic drug that may impede the progression of Alzheimer's disease,[†] presumably by interfering with inflammatory reactions to amyloid that further damage the brain. Clinical trials are presently being conducted to find out if these drugs really work. If they do, they will become an approved treatment for people at high risk of Alzheimer's disease, as well as for those who already have it.

::

The great success of studies of the DNA of families with Alzheimer's disease has stimulated studies of the DNA of families with the other mental illnesses, such as schizophrenia and mood disorders. None of these families are as riddled with these disorders as those families with early-onset Alzheimer's disease. Nevertheless, the increased concentration of certain mental disorders in families is impressive,[†] suggesting that they too have some genetic basis.

Schizophrenia is a good example. Children of a parent with schizophrenia, each of whom has half of that parent's DNA, have about an 11 percent risk of becoming schizophrenic. So too do siblings of a person with schizophrenia, who also share half of the DNA of the affected person. Considering that everyone else's risk of being schizophrenic is only 1 percent, the elevenfold-greater risk to close relatives implies that gene variants shared by the affected family members play a part in the development of this disorder.

Another good example is bipolar disorder. As with schizophrenia, people with a parent or a sibling who has bipolar disorder have a substantial risk of being affected, in this case about 7 percent. And, as with schizophrenia, this is about ten times greater than the general risk of having bipolar disorder, again suggesting that gene variants play a role.

For the other mental disorders that I have considered, the relative risk to children and siblings of an affected person is also significant, but not as great. For ADHD it is about six times greater. For obsessive-compulsive disorder or panic disorder it is about five times greater. For major depression it is also about five times greater for children and siblings of those who became depressed before the age of twenty, although it is much less if the onset was in middle age.

Not long ago studies of such families sparked explosive debates between those who took them as evidence for learned familial patterns of abnormal behavior and those who took them as evidence of an inherited predisposition to a particular mental disorder. Now there is general agreement that environment and heredity both play some part. There is also general agreement that there is much to be gained from the examination of the DNA of members of these families.

As with Alzheimer's disease, two approaches are being used to find the gene variants that play a part in the development of mental disorders.[†] The first, called the candidate gene approach, looks for particular variants of a certain gene in people with a mental disorder. The gene selected for examination, called a candidate gene, is chosen on the basis of current ideas about the pathogenesis of the disorder. For example, in the case of Alzheimer's disease, the gene for APP seemed like a reasonable candidate because variations in its structure could influence the accumulation of A-beta and of amyloid, and careful examination of the APP gene turned up the London gene variant, which can indeed cause Alzheimer's disease.

The second approach, called the linkage approach, compares the DNA of the family members who have a particular disease with the DNA of those who don't, in the hope of finding the critical genetic difference between them. The linkage approach is based on the fact that each human gene is located at a specific position on one of twenty-four different structures called chromosomes, whose essential ingredients are chains of DNA. By correlating the inheritance of a disease with the inheritance of gene variants, called markers, that have known locations on particular chromosomes, the gene variant responsible for the disease can be located and identified because it is physically linked to the markers on a chain of DNA.

Although the linkage approach is much more expensive and time-consuming than the candidate gene approach, it has the great advantage that it does not depend on any preconceptions. In patients with early-onset Alzheimer's disease, application of the linkage approach led to the identification of previously unknown variants of the genes for presenilin-1 (on chromosome 14) and presenilin-2 (on chromosome 1). The same approach also helped to identify APOE-4 as a susceptibility gene variant (on chromosome 19) in the late-onset form of the disease. These unanticipated findings greatly increased our understanding of the nature of Alzheimer's disease, and, importantly, suggested some targets for drug development.

In the case of the mental disorders that are the main subjects of this book, the candidate gene approach has had limited success so far. One gene that is presently receiving a great deal of attention encodes an enzyme called COMT. Discovered more than half a century ago by Julius Axelrod, this enzyme, like monoamine oxidase, terminates the actions of dopamine and norepinephrine. In humans there are variants of the COMT gene that encode different forms of the enzyme. Recent evidence indicates that people with schizophrenia are more likely to have a particular form of the enzyme,[†] which raises the possibility that differences in the inactivation of dopamine or norepinephrine may play a part in the development of this disease.

The candidate gene approach has also been applied to attention deficit hyperactivity disorder. Because the main drugs for ADHD increase neurotransmission by dopamine, the dopamine transporter and the five dopamine receptors seemed like reasonable candidates. Several studies have indeed found that a particular variant of the gene for dopamine receptor-4 (D4) is more common in people with ADHD. Several others have found that a particular variant of the dopamine transporter gene is more common in people with this same disorder. Presently a group of investigators, who are organized as the ADHD Molecular Genetics Network, is studying the combined effects of these two gene variants on the risk of ADHD.[†]

The linkage approach has also had some limited success in identifying shared regions of DNA in affected members of families with many mental disorders. The most progress has been made in linkage studies of families with schizophrenia or bipolar disorder, the two disorders that appear to be the most heritable. In each case a number of DNA regions have been located that probably contain gene variants that influence the risk of the disorder. A search is under way for the relevant genes.

One reason that progress has been so slow is that, unlike early-onset Alzheimer's disease, no one has yet found cases of schizophrenia or bipolar disorder that are caused by the inheritance of a variant of a single gene. Instead, the risk of developing these mental illnesses is determined by the combined actions of multiple susceptibility gene variants acting in concert with environmental factors. The involvement of multiple genes and their complex interactions with each other and with the environment is the reason these disorders are called multiple gene disorders (or complex genetic disorders), a category that includes late-onset Alzheimer's disease. In the case of schizophrenia and bipolar disorder, none of the relevant gene variants may be as influential as APOE-4 is in late-onset Alzheimer's disease, making their detection particularly difficult.

The other reason that detection of such gene variants may be so difficult is that seemingly identical cases of a particular disorder may have different genetic origins. We have already encountered this sort of thing—called genetic heterogeneity—in the discussion of early-onset Alzheimer's disease, which can be caused by a gene variant of APP, presenilin-1, or presenilin-2. What makes genetic heterogeneity so challenging for gene hunters is that DNA studies of a mixture of cases of the three different types would have precluded the identification of any of these gene variants as causes of early-onset Alzheimer's disease. Such a study would not have identified a *shared* gene variant in all the affected people, because any of three different gene variants could be responsible. One reason why the Alzheimer's disease gene hunt succeeded in the face of such genetic heterogeneity is that the first successful studies restricted their attention to large families whose members all had the identical gene variant—such as the original London variant of the gene for APP. Another reason is that some of the studies were confined to members of a particular ethnic group called Volga Germans, who are more likely to share gene variants than a random selection of people. It was the studies of early-onset Alzheimer's disease in families of Volga Germans that resulted in the identification of the gene variant of presenilin-2.

For disorders such as schizophrenia, which are influenced by the combined actions of multiple susceptibility genes, the likelihood that different *combinations* of gene variants may be involved in different patients further complicates the search for each of them. Consider, for example, a situation in which variants of six different genes all contribute to the risk of schizophrenia, and inheritance of a combination of any three of them is sufficient to make it likely that a person will develop schizophrenia. In examining the DNA of people with schizophrenia, no gene variant would be found in all of them, making it difficult to implicate any gene variant in the pathogenesis of this mental disorder.

Unfortunately, everything we know about the inheritance of mental disorders such as schizophrenia suggests that multiple gene variants—perhaps dozens—are involved, and that inheritance of combinations of quite a few of them may be necessary to produce a high risk of developing the characteristic symptoms. Furthermore, the tactic of restricting attention to single large families cannot be successfully employed in the hunt for gene variants that contribute to a complex genetic disorder, because such hunts require the study of many more patients than can be obtained from even the largest families.

But, as with studies of the Volga Germans, which led to a gene variant for Alzheimer's disease, studies of ethnic groups from restricted geographic locations are facilitating the search for the genetic basis of schizophrenia. A notable example is Iceland, whose 278,000 people are descended from a small number of Norwegian Viking males and the Celtic women they brought with them. Many Icelanders are now participating in an ambitious study of their gene variants that is being conducted by DeCODE Genetics, an Icelandic biotechnology company. Scientists from DeCODE have recently published evidence that implicates a variant of a gene called neuregulin-1[†] in about 15 percent of Icelandic patients with schizophrenia. Furthermore, work with Irish schizophrenic patients by a group led by Richard Straub and Kenneth Kendler has implicated a variant of a gene called dysbindin.[†] Although the roles of neuregulin-1 and dysbindin in schizophrenia are not yet established, this work has generated a great deal of excitement because both are believed to participate in the formation of connections between nerve cells. This implies that differences in the physical connections themselves—and not just differences in the mechanisms of neurotransmission that they use to communicate—may play a role in the development of this disorder.

::

In addition to these encouraging results the main reason to be optimistic about gene hunts is that new technologies keep making it easier to examine samples of DNA. These technologies have already been used in a worldwide program, called the Human Genome Project, to discover a large part of the sequence of the three billion nucleotide building blocks in a sample of human DNA. Now this technology is being used to catalogue the three million common variations in these three billion nucleotides, which account for much of the inherited human diversity, including individual susceptibility to particular mental disorders. So powerful are these new technologies that several biotechnology companies are competing to devise schemes that would catalogue all the variations in any person's DNA and then compare them with other samples using a high-speed computer. As this technology is developed, and as the costs keep coming down, it will be possible to identify all the gene variants[†] in each of a very large number of people with a disorder such as schizophrenia, to find those combinations of variants that are correlated with the disorder.

Another tactical change that will facilitate gene discovery is to simplify the problem by studying the genetic basis for behavioral traits that increase the risk of a mental disorder, rather than by focusing only on the full-blown pattern of symptoms. For example, there is evidence that people with schizophrenia and some of their close relatives share certain functional differences in a brain region called the prefrontal cortex. This brain region, which is located beneath the forehead, plays an important role in integrating complex mental processes, and these functional differences may contribute to the development of schizophrenia. Such differences can be detected with standardized psychological tests. They can also be detected by observing the activity of the prefrontal cortex with a brain scanning method called functional magnetic resonance imaging. Because these differences are fairly easy to detect, scientists have been examining the function of the prefrontal cortex in all the relatives of people with schizophrenia, in the hope of correlating functional differences with particular gene variants. Recent studies suggest that a variant of the COMT gene is indeed more common not only in people with schizophrenia but also in their relatives with functional abnormalities of the prefrontal cortex.[†] This supports the inference that it is a susceptibility gene variant for schizophrenia.

The observation that some people who are not mentally ill share certain psychological, physiological, and genetic differences from the norm with their schizophrenic relatives is also leading to a rethinking of the classification of psychiatric disorders. As I have already pointed out, these disorders are presently classified categorically. This means that if you fulfill certain criteria, you are put in this category. For example, schizophrenia, as presently defined, is a categorical diagnosis—you either have it or you don't. The alternative to this black-and-white approach is to use dimensional diagnoses. These are based on quantification of various attributes in various shades of gray and allow for the classification of some individuals into intermediate behavioral categories. If the risk of schizophrenia is influenced by the cumulative effect of multiple gene variants, those people in the intermediate behavioral categories might have an intermediate number of these variants. Using a dimensional approach may be necessary to find the gene variants that influence susceptibility to this and other mental disorders.[†]

As the relevant gene variants are found, they will provide a basis for studying the pathogenesis of mental disorders. But figuring out the many biological effects of particular gene variants may take some time. For example, in the case of the presenilins, it took several years to figure out that they are components of an enzyme that cleaves APP. In the case of APOE-4, it is still not clear whether it contributes to the pathogenesis of Alzheimer's disease by augmenting the accumulation of amyloid or by some other means. As its role is discovered, this information will guide the development of new drugs.

Even then, creating new drugs may prove to be very difficult. Knowing that a biological process has gone awry as a result of the actions of a variant protein is no guarantee that a drug can be found that will fix it. In the case of Alzheimer's disease there is great hope that the secretase inhibitors developed on the basis of genetic discoveries will turn out to prevent A-beta accumulation in human brains, and stop the progression of the nerve cell dysfunction and death. But this approach may not work. Even if it does, the side effects of the secretase inhibitors may be so severe that no one would be willing to take them. Already there are concerns that other normal proteins are cut up by gamma secretase, one of the enzymes that makes A-beta, and that blocking this secretase may interfere with vital functions controlled by those protein fragments. So the price of reducing the formation of A-beta with a blocker of gamma secretase may turn out to be unacceptable.

Fortunately the growth of genetic technologies is providing new ways to create and evaluate drugs for mental disorders without resorting to time-consuming, expensive, and potentially dangerous testing on patients. The trick, as you will now see, is to adapt these technologies to create genetically modified animals that can be used in their place.

8 :: *Animal Psychiatry*

Go to the ant, thou sluggard; consider her ways, and be wise.
—Proverbs 6:6

The idea that animals can be used to study mental illness strikes many people as strange, because human behavior seems unique. After all, only humans use language for introspection and long-range planning, and it is just these functions that are disturbed in many psychiatric disorders. Nevertheless, we have enough in common with other animals to make them very useful for studies that can't be done with patients.

Of the animals used for this purpose, apes and monkeys have been favorites because they are our closest relatives. Dogs, too, have obvious human qualities. They may even display patterns of maladaptive behavior that resemble those in *DSM-IV*. For example, Karen Overall, a professor in the School of Veterinary Medicine at the University of Pennsylvania, has been studying a dog version of obsessive-compulsive disorder (OCD),[†] which is fairly common in certain breeds. Like people with OCD who each have their particular patterns of symptoms, individual dogs with canine OCD also have distinctive main symptoms such as tail-chasing or compulsive licking of their paws. Like humans with OCD, the dogs tend to perform their rituals in private. And, like human OCD, the canine version responds to drugs[†] such as clomipramine and Prozac. Because of these many similarities, Overall's dogs may provide information about the human disorder—an aim that has already been achieved for another canine behavioral disorder that I will turn to shortly.

But despite their value for certain types of studies, primates and dogs are not ideal experimental animals. Their main shortcoming is that they are costly to raise and maintain, which makes them impractical for the many experiments that require large numbers of subjects. For this reason scientists have been turning to a much less expensive alternative, the laboratory mouse. Although it is more difficult to empathize with these tiny rodents than with a chimpanzee or a golden retriever, we now know that all these mammals share much of our complex brain machinery. What makes mice especially attractive is that their genes are relatively easy to manipulate by traditional breeding methods and by the new techniques of genetic engineering. Both experimental approaches have been successfully employed to make special strains of mice that are being used to study mental disorders and to develop new psychiatric drugs.

::

Breeding mice for biomedical research began early in the twentieth century. At first this was done by people in the pet business, who had learned how to raise unusual strains with pretty coats. As the demand from scientists grew, academic institutions such as the Jackson Laboratory in Bar Harbor, Maine, were established to produce large numbers of mice under carefully controlled conditions. Nowadays nonprofit and commercial breeders are raising millions of these animals every year. Millions more are being bred in the laboratories of individual researchers.

The main contribution of the professional mouse breeders is to prepare and maintain the uniform populations, called inbred strains, which scientists prefer to use for their experiments. Inbred strains of mice are relatively easy to make because these animals reproduce and mature so quickly. Pregnancy takes only three weeks, and newborn mice become sexually mature in about six weeks, so the interval from one generation to the next can be as brief as a few months.

To establish a particular inbred strain, a founder male and

female are mated, and a daughter and son from this first generation are chosen as parents for the next generation. When these siblings are mated, their offspring will be more genetically alike than their parents. Repeating such brother-sister mating for generation after generation produces siblings who are progressively more alike—because in each generation some of the parents' DNA is not passed on. After two dozen generations, which can be raised in as little as five or six years, all the members of the inbred strain have very similar DNA. As the inbreeding continues, the offspring become virtually identical.

The reason scientists prefer inbred strains for their research is that using them minimizes the innate variations between individual animals that can complicate the interpretation of experiments. For example, if an animal that receives a drug responds differently than an animal that receives an inert substance (a placebo), this difference could be due either to an effect of the drug or to a genetic difference between the two animals. But if both animals are the genetically identical members of an inbred strain, the difference is much more likely to be due to the drug. In practice, comparisons are generally made between groups of animals rather than individuals, thereby increasing confidence in the reliability of observed differences. Nevertheless, using members of an inbred strain reduces the number of animals that must be compared to get a reliable result, because there is so little innate difference between them. Furthermore, widespread use of a particular inbred strain makes it much easier to compare the results of experiments done by scientists in different laboratories around the world.

::

In the course of the twentieth century several inbred strains of mice became popular for general laboratory use. Two examples are BALB/c, an albino strain, and C57BL/6, a black strain. As experience was gained with these two strains, it became apparent that they have different personalities. Some of the differences can be measured with simple psychological tests that can be performed in just a few minutes.

One of the most widely used mouse psychological tests, called the open-field test, measures fearfulness. In this test a mouse is placed in a brightly lit box that is open at the top. In this exposed position the average mouse shows signs of fearfulness such as defecation, urination, decreased movement (freezing), and huddling along the walls of the box. The degree of fearfulness can be estimated by measurement of the mouse's exploration of the box with motion detectors: greater fearfulness translates into less exploratory activity.

In a closely related test, called the elevated-plus-maze test, the animal is placed in an apparatus that is raised about fifteen inches above the ground, a considerable height for a mouse. The apparatus has four arms, arranged like a plus sign. Two of the arms have black walls that create a protected place, whereas the other two arms have no walls. Fearfulness is estimated by the time spent in the protected arms as opposed to the relatively dangerous open arms. In another related test, called the light-dark box test, a mouse is placed in a dark box that has an open doorway to a brightly lit box, and fearfulness is estimated by the time the mouse spends in the dark box. What makes these simple tests so informative is that there is evidence that the mouse brain circuits and neurotransmitters that control the fearfulness that they measure are similar to the brain circuits and neurotransmitters that control fearfulness in people.

When observed in the open field test, a typical BALB/c mouse behaves differently than a typical C57BL/6 mouse. Although neither strain was deliberately bred for either fearfulness or fearlessness, the albino mice are only about 20 percent as active as the black mice in this stressful environment. They are also more likely to defecate and urinate when placed in the open field, which is another indication that they are more afraid. These findings suggest that members of the albino strain are innately more fearful than those of the black strain because of differences in certain gene variants. This interpretation is supported by the results of breeding BALB/c mice with C57BL/6 mice and further breeding of the mixed offspring. The grandchildren and great-grandchildren, who have mixtures of

BALB/c and C57BL/6 gene variants, have open-field behavior that is intermediate between those of the two ancestral strains.

Based on this and other evidence that heredity influences the fearfulness of mice, John DeFries and colleagues at the University of Colorado set out to breed lines of mice to find out how fearful or fearless they would become.[†] Starting with great-grandchildren derived by breeding of BALB/c and C57BL/6 mice, they mated the most fearless male and female as well as the most fearful male and female. Such selective mating was continued for generation after generation.

The results were striking. After thirty generations the mice that were bred to be fearless were about thirty times as active in the open field as those bred to be fearful. The difference between them developed gradually over the course of many generations of selective breeding. This indicates that multiple gene variants influence this behavioral trait, and that certain of these variants become progressively more prevalent as the selection process continues.

The reason fearful and fearless groups of mice can be developed in this way is that many gene variants that influence these aspects of behavior already exist in the mouse population. Each gene variant arose in the distant past by random changes—called spontaneous mutations—in the DNA of an egg or sperm of an individual ancestral mouse. Many of the gene variants that are found in contemporary mice were retained because each has a beneficial effect under certain circumstances. In the particular natural state in which mice find themselves, a balance is reached between gene variants that favor innate fearfulness and those that favor innate fearlessness, because both these propensities have benefits. Were mice to become less fearful, they might explore their environment more freely and find additional food—but at the cost of greater exposure to predators. Were they to become more fearful, the balance would shift in the opposite direction. Based on these and other natural selective forces, mice acquire a particular balance of gene variants that promote fearfulness and gene variants that promote fearlessness, by whatever biological effects these gene variants achieve this behavioral result.

Changes in the prevalence of predators and in the availability of food lead to a shift in the balance of these gene variants.

Creation of extremely fearful and fearless lines of mice by deliberate breeding replaces the natural selective forces with those imposed by the breeder. The experiment is important not only as a way of proving the existence of gene variants that influence these emotional states, but also as a step in identifying the particular gene variants that are responsible. By correlating the behavior of inbred strains of fearful and fearless mice with detailed studies of their DNA, scientists have identified the chromosomal locations of the genes that influence these patterns of behavior,[†] and are beginning to track down the relevant gene variants.

This, then, is one simple way to use mice to try to find the origins of a psychiatric problem, in this case the inherited vulnerability to develop pathological forms of anxiety. Although the gene variants that control the propensity to fearfulness in mice may turn out to be different from those in humans, it is a good bet that the discoveries in mice will contribute to an understanding of the human condition. The reason to be optimistic is that it seems likely that the mouse gene variants that influence fearfulness will lead to the identification of a network of genes and proteins that work together in all mammals to regulate this critical emotional pattern. This may open up a new approach to the pharmacological treatment of anxiety disorders.

::

Although selective breeding of mice is beginning to provide us with information about the genetic basis of behavior, it can only tell us about the functions of the gene variants that mice already possess. To increase the usefulness of mice for psychiatric research, techniques have been developed to create animals that have new gene variants[†]—variants not already found in the mouse genetic repertoire. In some cases this has been successfully done by deliberately creating random mutations in the DNA of thousands of mice,[†] with a chemical called ENU, and then selecting those mutants with interesting patterns of behavior. But the most popular way to introduce new gene variants is by a more selective manipulation of mouse DNA with a technology called genetic engineering.

To make a genetically engineered mouse, the first step is to transfer a particular gene into the DNA of a mouse embryo. In the simplest case the gene is inserted into a random location in the DNA chain of a chromosome. As the embryo grows by cell division the added gene is copied along with the rest of its genes, and becomes a permanent part of the genetic makeup of each cell. When the modified embryo has developed into an adult, its egg cells or sperm cells contain the transferred gene, which is transmitted to its children along with the rest of its genes. These children then transmit it to their children. The end result is a line of mice called transgenic mice because they contain the transferred gene.

In some cases the gene that is transferred to a transgenic mouse comes from a human being. A notable example is the transfer of the London or Swedish variants of the APP gene, each of which came from the DNA of a person with Alzheimer's disease, and each of which gives rise to a form of Alzheimer's disease in the mice that receive it.[†] The first sign that the transferred gene has transmitted the disease is the appearance of memory impairment, as measured with standard mouse tests. Failing memory is detectable when the transgenic mice are about twelve months old—the equivalent of human middle age—and becomes more pronounced in the ensuing months.

Human APP gene variants are not the only ones that bring on mouse dementia. Transfer of a human presenilin-1 (PS1) gene variant derived from a person with early-onset Alzheimer's disease speeds up the development of memory impairment in mice that already contain the London or Swedish variant of the APP gene. In lines of transgenic mice with both an APP gene variant and a PS1 gene variant,[†] evidence of Alzheimer's disease becomes apparent when they are only six months old.

Mice with these transferred genes are being used to study the pathogenesis of Alzheimer's disease—the process by which accumulation of A-beta leads to the dysfunction of nerve cells. Although the brain changes in these mice are not identical with those observed in the human disease, they are sufficiently similar to be informative. The mice are also being used to test drugs, such as inhibitors of beta-secretase and gamma-secretase, to determine whether they can prevent or interrupt this disease process.

As the gene variants that influence the susceptibility to disorders such as schizophrenia are identified, they too will be transferred into mice to create transgenic lines. This approach, which is working so well for studies of the pathogenesis and treatment of Alzheimer's disease, may also prove useful for many mental disorders.

::

Genetic engineering of mice is done not only by inserting a gene into a random position in mouse DNA. There is another technique, called gene targeting, in which an existing gene variant (variant 1) is replaced by an alternative variant (variant 2) of the same gene. With this technique variant 1 is cut out of its normal position on a chromosome and variant 2 is substituted in exactly the same location. The end result is a line of mice with variant 2 instead of variant 1.

In most cases variant 2 is a man-made gene variant that has been prepared in a test tube by chemical manipulation of isolated copies of variant 1. The chemical manipulation is designed to make a specific change in the function of variant 1 by substituting or removing some of its nucleotides. In many cases variant 2 is designed with a major flaw that makes it nonfunctional, then inserted into the DNA of embryos by a technique that exchanges it for variant 1. The descendants of embryos whose variant 1 was replaced by a nonfunctional variant 2 are called "knockout" mice, because the function of the gene has been knocked out.

Knockout mice are generally used to find out what a gene normally does, by observing the effects of its functional elimination. But this is not as simple as it sounds because the gene may be involved in a number of interacting biological processes. Furthermore, the body has ways of compensating for the loss of certain genes by increasing or decreasing the activities of related genes. Nevertheless, knockout mice are providing a great deal of information about the roles of genes in biological and behavioral processes and in the pathophysiology of many diseases.

Some of the information from knockouts has been very surprising. For example, a notable surprise came from the knockout of a gene that encodes a protein that can be chopped up to produce two different brain peptides. Called hypocretins because they are made in the hypothalamus, these two peptides are also sometimes called orexins (from the Greek word for "appetite"—the same root that is used in "anorexia"), because they are believed to increase appetite. I will refer to them as hypocretins.

To study the behavioral functions of the hypocretins, Masashi Yanagisawa and his colleagues at the University of Texas knocked out the hypocretin gene, thereby deleting both of these peptides. When they videotaped the knockout mice to monitor their movements, they were amazed to find that the mice kept falling asleep for short periods.[†] This abnormal pattern of behavior has all the characteristic features of narcolepsy, the human disorder that I first mentioned in discussing the early uses of amphetamine.

While these observations were being made, a role for hypocretins in narcolepsy was discovered independently by Emanuel Mignot and his colleagues at Stanford University. Mignot had been studying a genetic form of narcolepsy in a line of Doberman pinschers. Using the same gene-hunting techniques that successfully led to the identification of gene variants that cause Alzheimer's disease, he searched the DNA of the narcoleptic dogs and located the gene variant that distinguished them from their relatives who do not have the disease. He was astonished to find that this variant encodes an inactive form of a receptor for one of the hypocretins[†] (hypocretin receptor-2). Then, based on Mignot's discovery, Yanagisawa's group knocked out this hypocretin receptor in mice and confirmed that its absence—like the absence of the hypocretins themselves—causes narcolepsy.

As soon as these discoveries were made, attention shifted to human narcolepsy. But unlike narcolepsy in Dobermans, humans with narcolepsy have normal genes for hypocretin receptors. They also have a normal hypocretin gene. So what does the hypocretin system have to do with the human disease? Because the animal studies clearly implicated hypocretins in the pathophysiology of narcolepsy, several groups of scientists took a close look at the levels of these peptides in human brains.

They were rewarded with the discovery that the brains of people with narcolepsy do indeed have abnormally small amounts of hypocretins,[†] but not because of an inherent inability to make them. Instead, the hypocretin deficiency in human narcolepsy is due to an autoimmune reaction[†]—the same type of disease mechanism responsible for both Graves' disease and the autoimmune thyroxine deficiency that frequently follows in its wake. Just as autoimmune thyroxine deficiency is caused by the destruction of thyroid cells by specific antibodies, so too is the hypocretin deficiency that causes narcolepsy due to the destruction of the nerve cells that make hypocretins by another group of specific antibodies. Although the reasons for these abnormal reactions of the immune system are not yet known, both of them appear to be due to an inherited predisposition to make certain self-destructive antibodies, combined with unidentified environmental factors.

Fortunately, the thyroxine deficiency that results from thyroid cell destruction can be treated by replacing the missing hormone with thyroxine pills. The same replacement approach should also work for human narcolepsy. But unlike thyroxine, which works if taken by mouth because it is not affected by intestinal enzymes, hypocretins taken by mouth are destroyed by the intestinal enzymes that we rely on to digest the proteins in our diet. For this reason hypocretins that are administered as pills can't make it to the brain to eliminate the repeated episodes of sleepiness. To get around this problem, it will be necessary to create artificial chemicals that mimic hypocretins but are not destroyed in the intestine. So even though research on mice and dogs has clarified the pathophysiology of human narcolepsy, it may take many years to make a substitute for hypocretins that is clinically useful.

For now the only treatments for narcolepsy are drugs whose effectiveness was discovered by trial and error. The old standbys are stimulants such as amphetamine. There is also a relatively new drug for narcolepsy called modafinil (Provigil), discovered in the 1980s,[†] which promotes wakefulness without directly affecting the actions of dopamine or norepinephrine. Although the interactions between these drugs and the hypocretin system have not yet been worked out, there is evidence that both amphetamine and modafinil influence the nerve cells that make these peptides.

::

The unexpected results of knocking out the hypocretin gene are only one example of the usefulness of gene targeting. Hundreds of other brain genes have been knocked out in mice, and many of these experiments have also been very informative. Nevertheless, the crude all-or-none nature of knocking out genes frequently results in gross abnormalities that make it difficult to interpret such experiments. To get around this limitation, new techniques are available for reversibly knocking out genes and for selectively knocking them out in particular brain regions.[†]

As more is learned about brain genes, gene targeting is also being used in a more subtle way. For example, instead of knocking out an existing gene variant, a new variant is engineered that inserts (or "knocks in") a particular feature, without altering the other functions of the gene. Lines of mice that have been engineered in this way are called "knock-in" mice.

Knock-in mice are being used for many purposes. Among them is the design of new psychiatric drugs. For example, knock-in mice are being used to help search for a hitherto elusive medication—a drug that relieves anxiety without producing sleepiness. Because this search is succeeding and is such a good example of this approach, I will give you some of the details.

The search for an antianxiety drug that doesn't make you sleepy goes back to the early part of the twentieth century, when highly sedating drugs such as phenobarbital were the leading treatment for this common form of mental distress. A major advance in this search came more than forty years ago, when it was discovered that benzodiazepines such as diazepam (Valium) relieve anxiety while causing less sleepiness than phenobarbital. But Valium and other benzodi-azepines are still somewhat sedating. In fact, many short-acting ben-zodiazepines are widely used as sleeping pills.

To address this limitation, pharmaceutical companies kept trying to find a nonsedating benzodiazepine. Accustomed to a trial-and-error approach, they made thousands of chemically modified benzo-diazepines and tested their effects on anxiety and sedation in exper-imental animals. They were rewarded with a few new drugs, such as alprazolam (Xanax) and clonazepam (Klonopin), which have some useful features. Unfortunately, all the new ones are also sedating.

Having exhausted this trial-and-error approach, attention shifted to the targets of benzodiazepines, called GABA-A receptors. At first these receptors were very hard to study, because there are many dif-ferent types that are mixed together in bits of brain tissue, and are difficult to distinguish. But studies of the genes for GABA-A recep-tors revealed that all these receptors are made by combining three main classes of component proteins, including one called alpha. Furthermore there are six different alpha proteins (alpha 1 to 6)—each encoded by a different gene. Four of them (alpha 1, 2, 3, and 5) bind benzodiazepines.

Once this was established, the next step was to find out if the antianxiety effect and the sedating effect of the drugs are controlled by binding to the same or different alpha proteins. One way to answer this question is to try to make drugs that bind only to a sin-gle alpha protein, and see how each of them affects fearfulness and sleepiness in animals. Another approach is to tamper with the mouse genes for each of the four alpha proteins that bind benzodi-azepines.

The gene-tampering approach is attractive because so much is now known about the binding of benzodiazepines to alpha proteins. All four of those that bind benzodiazepines have the amino acid his-tidine (H) at a critical position in their protein chain, whereas the two that don't have a different amino acid—arginine (R)—at this critical position. Furthermore, if the H is replaced with an R in any of the benzodiazepine-binding alpha proteins, the genetically engi-neered protein no longer binds benzodiazepine.

These observations immediately suggest a knock-in experiment—the creation of mice in which the gene for one of the Valium-binding alpha proteins is replaced by an inactive alternative. Once a knock-in mouse has been created, it can be tested with Valium. Does the Valium reduce anxiety, cause sedation, or both?

The first of these knock-in experiments was done by a group of scientists at the University of Zürich and the Swiss Institute of Technology in collaboration with colleagues at Hoffman–La Roche, the Swiss pharmaceutical company that first brought us benzodiazepines. They replaced the normal gene for the alpha-1 protein with a variant that doesn't bind benzodiazepines.[†] When these modified mice are given Valium, the drug binds to the alpha proteins that have not been tampered with. But it can't bind to the modified alpha-1 protein.

The result, which was reported in 1999, was dramatic. The Valium was still effective in relieving the fearfulness of the knock-in mice, as measured with an elevated maze and a light-dark box, despite the fact that it cannot bind to their slightly modified alpha-1 protein. But these very same mice were not sedated by Valium. The sedative effect and the antianxiety effect can be separated!

In less than a year, scientists at Merck's Neuroscience Research Center in England reached the same conclusion, not only by confirming the Swiss knock-in experiment but also by studying the effects of an experimental benzodiazepine in normal mice. This experimental drug, called L-838,417, can distinguish normal alpha proteins: it binds to the alpha-2 and alpha-3 proteins but not to alpha-1. As expected from the knock-in experiment, L838,417 does not produce sedation, because it doesn't bind the alpha-1 protein,[†] the one that is responsible for the sedative effect. But it does reduce fearfulness, as measured by several tests. This suggests that the antianxiety effect of benzodiazepines is due to binding to the alpha-2 protein, the alpha-3 protein, or both. Furthermore, it makes L838,417 the first example of that long-sought medication: a nonsedating benzodiazepine.

Not to be outdone, the Swiss group promptly reported more knock-in experiments that helped to clarify how the Merck drug works and to more precisely define the target for the antianxiety effect of benzodiazepines. They made two more lines of knock-in mice: one with a modified alpha-2 protein that can't bind Valium, and another with a modified alpha-3 protein that can't bind Valium. Both lines of mice were given Valium. Those with the modified alpha-2 protein remained fearful in the elevated maze and light-dark box, indicating that the anxiety-reducing effect depends on binding to alpha-2.[†] In contrast Valium reduced the fearfulness of the mice with the modified alpha-3 protein, indicating that the antianxiety effect does not depend on binding to this protein.

Now that the alpha-2 protein has been identified as the target for the anxiety-reducing effects of benzodiazepines, the race is on to develop a nonsedating antianxiety drug that can be brought to market. Merck's experimental drug, L838,417, is one obvious starting point. But it remains to be seen how effective it is in patients. As with all drugs, there is always the danger of side effects, including rare ones that turn up only after lengthy evaluation. At least three major pharmaceutical companies—Merck, Roche, and Pfizer—are starting to test candidates in humans, hoping to be the first to capture a share of what might become a huge market.

But even these drugs may not turn out to be the perfect benzodiazepines, because they too may have some of the undesirable features of those that are presently used. The main drawback is the potential for abuse. As the new benzodiazepines are introduced, it remains to be seen if elimination of the sedating effect that comes with binding to the alpha-1 protein will also make them less attractive to people who are prone to addiction. Should they retain their addictive property, the search for a nonaddictive version will become the new challenge for the benzodiazepine pharmacologists.

::

Exciting as this work with animals is, it is only the beginning. Now that the value of studies of mouse fearfulness and sleepiness has been established, new behavioral techniques are being developed to evaluate other aspects of mouse psychiatry. Although no one has yet produced mice that are unequivocally depressed or schizophrenic, attempts are being made to identify critical features of the pathophysiology of these human disorders and then replicate them in mice. To do this, scientists are combining the powerful genetic techniques that are already available with new ways of studying mouse behavior.

Mouse genetics, a major underpinning of this work, is also advancing at breakneck speed. Once the first draft of the sequence of all human DNA (the human genome) was published in 2000, many molecular biologists and geneticists turned their attention to another major international effort—the mouse genome project[†]— to determine the sequence of all mouse DNA. We already know that the mouse genome closely resembles its human counterpart, and that even the grouping of genes on particular chromosomes is very similar. As the details of the mouse genome are worked out they will greatly facilitate further work in mouse psychiatry, such as the identification of gene variants that influence mouse fearfulness.[†] Increased information about mouse genes will also assist in the analysis of the physiological effects of the human gene variants that influence susceptibility to mental disorders.

While these difficult tasks are gradually accomplished, discoveries about human gene variants are also finding more immediate applications in psychiatry, by explaining why some people have odd responses to psychiatric drugs. This line of research is the topic of the next chapter.

9 :: *Personalized Psychiatric Drugs*

If it were not for the great variability[†] among individuals medicine might as well be a science and not an art.

—William Osler (1892)

When Michael Adams-Conroy, a nine-year-old boy from Martins Creek, Pennsylvania, arrived at the emergency room of a nearby hospital in February 1995, he was already dead. As in other cases of mysterious death, the county coroner ordered an autopsy. Based on the results, a homicide investigation was begun.[†] Michael's adoptive parents were the leading suspects.

Homicide was considered because tests of Michael's blood showed that he had been poisoned by Prozac, the drug he had been taking for several years as a treatment for obsessive-compulsive disorder. Although Prozac is thought of as being very safe, because even large overdoses rarely produce fatal blood levels, Michael had more of the drug in his blood than had ever been seen before. The police speculated that Michael's parents might have deliberately given him a massive overdose.

Pleading innocence, the Adams-Conroys sought help from experts who specialize in toxic reactions to drugs. They immediately offered an alternative explanation: Michael might have accumulated Prozac in his blood not because of a malicious overdose but, instead, because he lacked the enzyme that metabolizes the drug and leads to its excretion. Such an enzyme deficiency would be particularly dangerous because Michael's psychiatrist had been prescribing unusually large doses of Prozac—up to five times the amount Clara is taking. The combination of large doses of the drug and the inability to metabolize it could explain Michael's lethal blood level.

To evaluate Michael's ability to metabolize Prozac, a sample of his tissue was sent to Floyd Sallee at the University of Cincinnati. Sallee tested the tissue for an enzyme called cytochrome-P450-2D6 (also called CYP2D6, or 2D6), which converts Prozac and many other drugs to a form that leads to their excretion. He found that Michael's 2D6 was defective. Together with the high doses of Prozac that Michael received, his abnormal 2D6 was responsible for the fatal buildup of the drug.

Based on this evidence, the homicide investigation was closed, and the Adams-Conroys filed a malpractice lawsuit against Michael's psychiatrist. They charged that the doctor had negligently prescribed excessive doses of Prozac and did not deal properly with signs of toxicity that culminated in the seizures that caused Michael's death. The case was settled out of court.

::

Like so much else about psychiatric drugs, the key discovery about their metabolism was made in the 1950s. At the time it was already known that drugs are inactivated by the body before their excretion in the urine. But very little was known about the details of this process.

Much of the pioneering work on drug metabolism was done at the National Institutes of Health by Julius Axelrod, whose subsequent research on neurotransmitters I mentioned earlier. In 1953 Axelrod began studying the ways that the body inactivates amphetamine,[†] the drug that Gordon Alles had invented more than two decades before. He found that the liver contains enzymes that modify parts of this drug molecule. Furthermore, the same liver enzymes can also modify a variety of other drugs that have strikingly different properties. This distinguishes these enzymes from many other enzymes that are extremely specialized and react with a single chemical.

We now call this group of enzymes drug-metabolizing enzymes. But despite their name, the main targets of these enzymes are not products of the pharmaceutical industry. Instead they are toxic natural chemicals found in terrestrial plants,[†] which animals first started eating when they emerged from the sea, about four hundred million years ago. As soon as animals encountered these plant chemicals, they began evolving enzymes to destroy them. The plants responded by evolving many new toxins to repel their animal predators; the animals fought back by evolving better enzymes that can inactivate just about anything the plants produce. When synthetic drugs such as amphetamine came along, they were an easy mark for this elaborate detoxification system.

About half a dozen enzymes play major roles in the metabolism of the commonly used psychiatric drugs.[†] Each of these proteins is a different version of cytochrome P-450 (CYP), and each is the product of a different gene. One of the most important is CYP2D6 (2D6), which metabolizes not only Prozac but also many other drugs including imipramine (the original tricyclic antidepressant), amitriptyline (the first drug I gave Clara), propanolol (the first drug I gave Martha), and chlorpromazine (the original antipsychotic). Another, called CYP3A4, is the most abundant of these enzymes and also has many targets, such as alprazolam (Xanax) and several atypical antipsychotics. Many drugs are metabolized by more than one enzyme.

The interaction between drugs and these enzymes can be complex. Some drugs are not only metabolized by the enzymes but may also partially inactivate them. For example, Paxil and Prozac, two of the most popular SSRIs, substantially reduce the activity of 2D6 and can prolong the action of other drugs that are metabolized by this enzyme. Nefazodone (Serzone), another antidepressant, reduces the activity of 3A4 and prolongs the action of alprazolam (Xanax).

Drugs may also influence the activities of drug-metabolizing enzymes by stimulating the body to make more of them. This is also a common reaction to chemicals in herbal remedies. For example, St. John's wort, which is sold over the counter as an herbal remedy for depression, stimulates the body to make more of the CYP3A enzymes.[†] The increased manufacture of these enzymes is an adaptive response to the ingestion of the plant chemicals and helps the body get rid of them. But the accrual of CYP3A enzymes also accelerates the metabolism of other foreign chemicals such as ethinyl estradiol, a common ingredient in birth control pills. As CYP3A enzymes accumulate with continued use of the herbal remedy, more of the contraceptive is inactivated and becomes ineffective. For this reason women who take birth control pills are warned to avoid St. John's wort.

Doctors are generally aware of the combinations that cause problems, and often adjust doses or substitute other drugs to minimize drug interactions. In some cases the dose is adjusted after monitoring the amount of a drug in a patient's blood. Blood tests are especially important for those drugs that are toxic at blood levels that are not much greater than those needed to achieve a therapeutic effect. Blood tests may also provide the first indication that the patient has a defective drug-metabolizing enzyme, like the one that contributed to the death of Michael Adams-Conroy.

::

The discovery that some people have defective drug-metabolizing enzymes[†] was made in the 1970s by Robert Smith, a British pharmacologist. Intrigued by the observation that certain patients kept fainting after taking debrisoquine, which was then a popular medication for high blood pressure, Smith tried some himself. He was surprised to find that he, too, became light-headed after taking a standard dose of the drug, and that his blood pressure plummeted. In contrast, several of his colleagues who took the same dose of debrisoquine did not react in this way, raising the possibility that they metabolized it differently than he did.

To test this possibility, Smith compared samples of his urine and that of his colleagues after they had taken the drug. He found that his urine contained only 5 percent as much of the metabolized and inactive form of debrisoquine as theirs. From this, Smith concluded that his idiosyncratic reaction to debrisoquine, like that of the patients who fainted, is due to an impaired ability to metabolize the drug to an inactive form, and the resultant accumulation of large amounts of the intact drug in the blood and tissues.

We now know that Robert Smith is a slow metabolizer of debrisoquine because of an inherited deficiency of 2D6 and that the reason for this deficiency is that Smith inherited defective 2D6 gene variants from both his father and his mother. We also know that defective 2D6 gene variants are remarkably common. At least 25 percent of Caucasians have a defect in one of their 2D6 genes,[†] and at least 5 percent have defects in both. People with just one normal 2D6 gene still make enough 2D6 enzyme to efficiently metabolize debrisoquine or Prozac. But those whose 2D6 genes are both abnormal are slow metabolizers of these drugs.

The slow metabolizers are not the only ones whose 2D6 genes cause abnormal responses to medications. Other people are at the opposite extreme. Called ultrarapid metabolizers, they have a 2D6 gene variant that manufactures unusually large amounts of the 2D6 enzyme. The plethora of enzyme metabolizes drugs so quickly that a large dose may be needed to get a therapeutic effect.

The gene variants responsible for differences in drug metabolism are not uniformly distributed throughout the human population, as Smith himself was among the first to observe. For example, 1 percent of Swedes and 29 percent of Ethiopians have the 2D6 gene variant that makes them ultrarapid metabolizers.[†] Among Chinese, slow metabolizers of 2D6 drugs are rarer than they are among Caucasians. On the other hand, Chinese are more likely to have low levels of CYP2C19, which is one of the enzymes that metabolizes drugs such as diazepam (Valium) and citalopram (Celexa).

Why are there such differences in the distribution of these gene variants in different parts of the world? Whenever geneticists find that more than one variant of a gene is fairly common (as in the case of fearfulness gene variants and fearlessness gene variants in mice), they assume that each variant may have both beneficial and detrimental effects. The advantage of having a gene variant that makes an active drug-metabolizing enzyme is that the enzyme degrades toxic chemicals, but the downside is that many of these enzymes also degrade some useful chemicals, such as certain hormones that the body makes. Because of this downside, people might be better off without a particular enzyme if the plants whose toxins it inactivates don't grow in their neighborhood, or if they simply learn to avoid them. Such geographic and cultural differences in the consumption of toxic plants may influence some of the ethnic variations in drug-metabolizing enzymes.

It is also possible that some of these ethnic variations have no evolutionary advantage, but are due, instead, to random genetic variations called random genetic drift. Consider, for example, the striking ethnic differences in aldehyde dehydrogenase, an enzyme that participates in the metabolism of alcohol. When alcohol is consumed, it is initially converted to acetaldehyde, a very toxic substance. It is the job of aldehyde dehydrogenase to dispose of the toxic acetaldehyde as soon as it is made. Yet a large percentage of Japanese and other Asians have inherited gene variants that produce defective aldehyde dehydrogenase[†]—variants that are much less common among Caucasians. This enzyme deficiency, which has no obvious benefit, is responsible for the well-known flushing that affects many Asians when they consume even small amounts of alcohol.

::

Idiosyncratic responses to a drug are caused not just by individual differences in the way the body handles it, which are called pharmacokinetic differences (*kinetics* refers to movement, in this case movement of a drug through the body). They are also due to individual differences in the way a drug affects the body, which are called pharmacodynamic differences (*dynamics* refers to force, in this case the force the drug exerts on the body). The outward signs of these two different types of idiosyncratic responses can be the same. For example, Robert Smith's light-headedness when he took debrisoquine could just as easily have been due to a greater sensitivity of receptors that respond to the drug (a pharmacodynamic difference) rather than to slower metabolism of the drug (a pharmacokinetic difference). It was the urine test, which showed poor metabolism, that settled the matter.

Presently, pharmacodynamic differences are harder to pin down than pharmacokinetic ones because they don't show up in simple tests of urine or blood. Nevertheless, it is well known that particular drugs produce distinctive idiosyncratic responses in a significant number of people who take them. In many cases the idiosyncratic responses are attributed to pharmacodynamic differences without direct evidence that they are responsible.

Sometimes a drug is blamed for tragedies that affect people who take them. For example, in June 2001 a federal jury in Cheyenne, Wyoming, ordered GlaxoSmithKline, the giant pharmaceutical company that makes Paxil (paroxetine), to pay $6.4 million to relatives of Donald Schell† because of horrible events that they attributed to an idiosyncratic pharmacodynamic response. Shortly before these events the sixty-year-old Schell, who was depressed, had been given a prescription for Paxil by his internist. After taking just two pills, Schell killed his wife, his daughter, his granddaughter, and then himself.

According to Andy Vickery, the lawyer for the plaintiffs, the jury made the award because it decided that the two Paxil pills were largely responsible for the deaths. Their decision was based on evidence that GlaxoSmithKline knew that a small number of people become violent after taking Paxil but failed to provide a warning. Although the expert witnesses for the company claimed that there was no persuasive evidence that Paxil caused Schell's bizarre behavior, this was challenged by the plaintiffs. Their expert witnesses testified that this SSRI, like others, could affect some people in this way, presumably because of a pharmacodynamic difference. The jury concluded that Paxil was 80 percent responsible and Donald Schell was only 20 percent responsible for the murder and suicide. Having decided to award $8 million in this case, the jury's assignment of 80 percent liability accounts for the $6.4 million the drug company was ordered to pay.

The company is appealing the verdict. Among their arguments is that Paxil is an effective treatment for depression and that the doctor who prescribed it for Schell had good reason to expect that the drug would help him. Furthermore, there was no way to anticipate that Paxil would make Schell murderous or suicidal. Had the doctor withheld the medication from Schell, he might have been tried and convicted of negligence.

This was, in fact, the outcome of another famous lawsuit about psychiatric drugs. It was initiated two decades years earlier by Rafael Osheroff,[†] then a forty-two-year-old physician, against Chestnut Lodge, a private psychiatric hospital in Maryland. Osheroff had suffered from episodes of depression for years. When his depression intensified in 1979, he was hospitalized at Chestnut Lodge, which at the time specialized in the use of psychological treatment for severe mental disorders and tended to avoid the use of psychiatric drugs.

Unfortunately, Chestnut Lodge's approach didn't work for Osheroff. In the seven months he was treated there he remained deeply depressed but received no medications. Frustrated by this lack of improvement, Osheroff's family had him transferred to Silver Hill Foundation, another private psychiatric hospital. There he was given an antidepressant and gradually recovered.

But Osheroff's prolonged depression had gravely damaged his personal and professional life. By the time he was discharged from Silver Hill his wife had left him, his partner had dissolved their joint medical practice, and the hospital where he worked had withdrawn his accreditation. Claiming that his life was ruined because Chestnut Lodge withheld treatment with an antidepressant, Osheroff sued. After an arbitration panel ruled in his favor, Chestnut Lodge agreed to pay damages, and the case was settled out of court.

Why did Osheroff have a favorable response to an antidepressant while Schell went berserk after taking two doses of a similar drug? Were the drugs really responsible for the recovery of one and the death of the other because of fundamental differences in the way they affected their brains? How might such differences be identified and used to predict individual reactions?

::

Variations in pharmacodynamic responses may be due to many types of individual differences. Presently, differences in the proteins with which a drug interacts are getting most of the attention, because a great deal has been learned about them. We now know the structures of the genes that make the main receptors, transporters, and enzymes that bind psychiatric drugs, and the structures of many variants of these genes. We also know that some of these variants affect the functions of these proteins[†] and their response to the drugs. Attempts are presently being made to find out how these variants influence a drug's therapeutic effects, as well as its side effects.

The serotonin transporter, which is the primary target of so many antidepressants, is a good example. Many variants of the serotonin transporter gene have been identified. The group of variants that have attracted most attention are those with variations in a region of the gene called the promoter, which controls the amount of the transporter that is made and displayed on nerve terminals. Some of these variants, called short variants, have repeated short stretches of DNA in the promoter region, whereas others, called long variants, have repeated long stretches of DNA in this region. Among Caucasians about 40 percent of the variants are short, whereas among Japanese about 60 percent of variants are short.

People with short variants make less of the transporter protein and transport serotonin more slowly into nerve terminals. This physiological difference appears to have behavioral consequences.[†] Early studies showed that people with at least one copy of the short variant are more likely to have a personality trait called neuroticism, which includes a relatively high level of anxiety. Several subsequent studies confirmed this association, though some have not.

Attention has since turned to the effects of these gene variants on responses to drugs that bind to the serotonin transporter. Scientists at Vita-Salute University in Milan have recently reported that depressed people who have a short variant of the serotonin transporter gene[†] don't get much improvement when treated with fluvoxamine (Luvox), an SSRI. They also found that when fluvoxamine is given along with pindolol, a drug that augments release of serotonin, the combination does relieve depression.

Variants of the serotonin transporter gene have also been studied in patients with bipolar disorder whose episodes of depression had been treated with SSRIs. Whereas many of these bipolar patients benefit from these drugs, others are shifted from depression into mania, which can have disastrous consequences. Scientists at the Centre for Addiction and Mental Health in Toronto recently found that bipolar patients who have SSRI-induced mania[†] are more likely to have two copies of the short variant of the serotonin transporter gene. Should this finding be confirmed, it could become the basis of a DNA test to guide drug treatment of patients with this mood disorder.

Variants of targets for other psychiatric drugs have also been correlated with susceptibility to major side effects. For example, a variant of the gene for the D3 dopamine receptor appears to be implicated in the development of tardive dyskinesia,[†] the chronic movement disorder that gradually affects some people after prolonged treatment with traditional antipsychotic drugs, such as haloperidol. The relationship between this gene variant and tardive dyskinesia was observed in independent studies by groups in Canada, Israel, and Norway. In addition to this pharmacodynamic difference there is evidence that the risk of tardive dyskinesia is also influenced by a pharmacokinetic difference in the metabolism of antipsychotic drugs due to inheritance of certain variants of the gene that makes the enzyme called CYP1A2.

::

Studies of the gene variants that influence pharmacokinetic and pharmacodynamic responses to psychiatric drugs are only in their infancy. Presently these studies are largely confined to a small number of genes that are known to play roles in drug metabolism and neurotransmission. But the range of these studies is beginning to expand as more gene variants are identified and as efficient new techniques are established for examining thousands of gene variants in an individual DNA sample. Several biotechnology companies are already trying to develop a method for identifying all the common gene variants in a person's DNA at an affordable price.

As this new technology becomes available it will help identify relationships between the gene variants that predict favorable therapeutic responses to drugs and those that influence the risk of developing mental disorders. The combined information about risk and potential response will lead to genetic tests that will help guide the treatment for each individual. The tests will also warn of potential dangers, like the one that contributed to the death of Michael Adams-Conroy. They will, in time, become a foundation for the gradual replacement of our one-size-fits-all medications[†] with a new pharmacopoeia of personalized psychiatric drugs.

10 :: *Little Things in Hand*

A little thing in hand is worth more than a great thing in prospect.
—Aesop, *The Fisher and the Little Fish*

While the major pharmaceutical companies wait to see what psychiatric drugs will come from discoveries about the origins of mental disorders, they are not idle. Aware of the difficulties of creating truly novel medications, they have continued to invest billions of dollars every year to make better versions of the existing ones. By concentrating on a limited number of drug targets of established value, they avoid the costly and risky preliminary research process that is needed to identify new ones.

Since the introduction of effective psychiatric drugs in the 1950s, tens of thousands of chemicals that influence neurotransmission have been created and examined. But only a few are significantly better than those that were discovered more than four decades ago. Nevertheless, pharmaceutical companies eagerly trumpet even the smallest apparent advantages of the new drugs they have developed. When they consider the advantages to be substantial, they flood psychiatric periodicals with multipage advertisements.

As I have been writing this book an advertising campaign that has occupied a particularly large number of pages in psychiatric periodicals was mounted by Wyeth, the Philadelphia company that made an early fortune in psychopharmacology by marketing meprobamate as Equanil. In the 1980s scientists at Wyeth found that a chemical called WY-45030 blocks the reuptake of both sero-tonin and norepinephrine,[†] raising the possibility that it could be used to treat depression. At the time pharmaceutical companies were eager to replace the original "dirty" drugs such as imipramine, which affects multiple transporters and receptors, with "clean" drugs such as Prozac, which selectively affects the serotonin transporter. From this perspective WY-45030's dual action—on both serotonin and norepinephrine—suggested that it might be unacceptably "dirty," especially because its effect on the norepinephrine trans-porter can cause undesirable elevations of blood pressure.

Eventually Wyeth realized that the dual action of WY-45030 could also be viewed as a blessing. After all, there was already sub-stantial evidence that blocking reuptake of either norepinephrine or serotonin can relieve depression, so blocking both with the same drug could well be advantageous. The only reason replacements had been sought for antidepressants such as imipramine, which also have this dual action, is that these early drugs are exceptionally "dirty" and cause many side effects through actions on receptors for acetylcholine and histamine. Because WY-45030 does not have those side effects, Wyeth began to promote their new drug as the first member of a promising new category, called selective serotonin and norepinephrine reuptake inhibitors (SNRIs).

After years of experimentation Wyeth established that WY-45030, now named venlafaxine (Effexor; Effexor XR in its extended-release version), is indeed an effective antidepressant. Furthermore, there is some evidence that certain depressed patients get more complete relief from Effexor than from SSRIs.[†] Using the Hamilton-D rating scale to measure symptoms of depression after eight weeks of treat-ment, 45 percent of patients treated with Effexor scored 7 or less, compared with 35 percent of patients treated with SSRIs. Because a score of 7 or less is taken as a sign of recovery from depression instead of just partial relief of symptoms, Wyeth's multipage ads for Effexor show people cavorting under the slogan "I got my playful-ness back."

The success of Effexor did not go unnoticed by Eli Lilly, the manufacturer of Prozac. Still reeling from the expiration of its patent on Prozac in 2001 and its replacement by a generic version of fluoxetine that was immediately marketed by a competitor, Lilly is now testing its own SNRIs. It is presently pinning its hopes on duloxetine[†] (Cymbalta), which, like Effexor, blocks both the serotonin and norepinephrine transporters.

Duloxetine has a relatively greater effect on norepinephrine than Effexor,[†] which may or may not be an advantage. But whatever the relative merits of Effexor and duloxetine are found to be, no one expects that they will radically change the treatment of depression, because neither of them is fundamentally different from the antidepressant drugs that were discovered half a century before. Although each may prove to have some significant benefits for individual patients, many people will not be helped by them or will find them unacceptable.

The same is true of the other new products that are now being heavily advertised. Of these, Pfizer's ziprasidone (Geodon), an atypical antipsychotic, has the important distinction that it does not appear to cause as much obesity as its competitors, because of a different balance of effects on neurotransmitter receptors. But despite this advantage it has not yet gained widespread acceptance. Concerta, another huge source of advertising revenue for psychiatric publications, is an extended-release form of methylphenidate (Ritalin). Its advantage, like that of other new competitors, is that children with ADHD can take it once a day, rather than in repeated doses. As its advertisements point out, "once daily Concerta lasts from home to homework." This simple modification of an established medication is bringing its manufacturer, Alza, a growing share of a substantial market. It has also stimulated the vigorous marketing of other extended-release forms of both methylphenidate and amphetamine.

::

The only thing that drug companies like better than copying or modifying existing drugs, is to capitalize on the off-label uses of the products they already sell. What makes this so attractive is that each of these drugs has already gone through a time-consuming and expensive procedure of testing[†] that has shown it to be both safe and effective for the treatment of a particular malady. Should physicians decide to prescribe the drug for some other purpose, the extra sales are extremely profitable for the manufacturer because they do not require additional research and development costs. Should this off-label use catch on, the manufacturer may be willing to support double-blind placebo-controlled studies to test the drug's effectiveness for this purpose,[†] in the hope that a favorable result would increase sales. Unfortunately manufacturers often turn a blind eye to off-label use, and are not interested in putting their drug to the test, for fear that the results may not be in their favor.

One drug that has successfully passed the test is Paxil (paroxetine), SmithKline Beecham's SSRI. Having introduced Paxil as an antidepressant after Lilly's Prozac and Pfizer's Zoloft were already being widely used, SmithKline needed a novel selling point to help market their drug. Upon learning that some psychiatrists were using Paxil off-label to treat patients with social phobia, SmithKline funded a randomized placebo-controlled double-blind study of the effectiveness of this treatment. Conducted at multiple centers in the United States and Canada, the results were published in the *Journal of the American Medical Association* in August 1998. The authors concluded that "paroxetine is an effective treatment for patients with generalized social phobia."[†] On the basis of this and other evidence the FDA approved the use of Paxil for this disorder in May 1999.

Once this new use for Paxil was officially approved, SmithKline helped fund an extensive campaign to increase public awareness about social phobia. Conducted in collaboration with nonprofit organizations, such as the Anxiety Disorders Association of America, the campaign called attention to the large number of people who have a marked and persistent fear of social situations, and to the benefits of treatment. Aided by a cover story by Joannie Schrof and Stacey Schultz in *U.S. News and World Report* on June 21, 1999 ("How shy is too shy?"), sales of Paxil soared. Soon Paxil joined Prozac and Zoloft on the list of the top ten selling drugs of all kinds—not just psychiatric drugs.

The campaign to promote awareness about social phobia did not escape public criticism. Little more than a month after the story in *U.S. News and World Report,* Michelle Cottle, a senior editor of *The New Republic,* responded with a major article. Called "Selling shyness: How doctors and drug companies created the 'social phobia' epidemic," the article questioned the claim of some psychiatric experts that one in eight Americans has social phobia, arguing that many are simply shy and don't need treatment. As Cottle put it, "trying to differentiate a 'mild' case of social phobia from 'normal' shyness is like trying to nail Jell-O to the wall. . . . The assertion that social phobia is not 'just shyness' suggests a bright dividing line that simply does not exist."

Nevertheless, Cottle acknowledges, "None of this suggests that some people aren't suffering. Certainly there are those who have a debilitating fear of social interaction. They rarely leave their homes. They cannot attend school or hold down a job. Dating is an impossibility." Cottle also admits that people like this may benefit from treatment with Paxil or other SSRIs. Although the prevalence of social phobia may well have been greatly exaggerated by setting excessively stringent criteria for defining normal social interactions, establishing this new use for drugs already on the market is a significant achievement.

::

While pharmaceutical companies continue to concentrate on medications that affect the actions of serotonin, norepinephrine, and dopamine, they have also been paying attention to other brain chemicals. Some of these are peptide neurotransmitters that are secreted by specialized groups of nerve cells that influence emotional behavior. As soon as receptors for these peptides were identified, they became potential targets for new psychiatric drugs.

One of the most-studied brain peptides is called substance P. Discovered in the 1930s in a powder made from extracts of brain (hence the name P, for powder), it was first shown to participate in the transmission of pain signals. We now know that it is a member of a family of peptides, called neurokinins or tachykinins, which are found in many brain regions. Like other neurotransmitters, neurokinins are made by specific groups of nerve cells. And, like other neurotransmitters, neurokinins signal certain nerve cells by binding to receptors on their surfaces. Substance P binds preferentially to a receptor called the neurokinin-1 (NK-1) receptor.

Once the NK-1 receptor was identified, drug companies became interested in finding chemicals that block its binding of substance P, which they hoped to use to make pain pills. They knew from the start that they shouldn't waste their time trying to design modified peptides that block the receptor, because if taken by mouth, the peptides would be destroyed by the protein-degrading enzymes in the intestines. Fortunately it is possible to make nonpeptide chemicals that bind to peptide receptors but are not broken down by intestinal juices. In fact, nature has already made its own nonpeptide chemicals, such as morphine, that bind to receptors for brain peptides such as endorphins and that are active if taken orally.

Attracted to this challenge, Merck established a program for the creation of nonpeptide drugs that bind NK-1 receptors. When they didn't pan out as pain relievers, Merck considered other applications. Because studies in experimental animals showed that substance P is concentrated in brain regions that participate in stress responses, Merck decided to find out if blocking the action of substance P with one of their new drugs, MK-869, would relieve reactions to stress, such as anxiety and depression.

The result of a placebo-controlled trial in patients with major depression was published in *Science* magazine on September 11, 1998, and was accompanied by considerable fanfare. The study, which was conducted at four different sites, showed that MK-869 was as good an antidepressant as paroxetine (Paxil).[†] After six weeks of treatment, 54 percent of patients who received MK-869 had a reduction of their Hamilton-D depression score of at least 50 percent, compared with 46 percent of those who received paroxetine and 28 percent of those who received a placebo. Furthermore, patients treated with MK-869 had a reduction in anxiety like that in those treated with paroxetine. But unlike many of the patients who were treated with paroxetine, those who received MK-869 did not have sexual side effects.

The 1998 report about MK-869 generated considerable excitement because it seemed to open a whole new approach to the treatment of depression. As the authors put it: "These findings provide clinical evidence that substance P antagonism represents a well-tolerated distinct mechanism for antidepressant activity." They closed their paper with this comment: "The possibility that alterations in substance P or the NK-1 receptor are primarily involved in the pathogenesis of depression requires further investigation, which may lead to a better understanding of the pathophysiology of this disease."

Unfortunately, a subsequent study of MK-869 in another group of depressed patients did not turn out as well as Merck had hoped. Although many patients got better while taking MK-869, a similar number got better while taking the placebo. Undeterred by this setback, Merck is continuing with this program of research.[†] Other pharmaceutical companies are also studying the behavioral effects of drugs that block the NK-1 receptor.

::

Substance P is only one of a number of brain peptides whose receptors are being studied by drug companies as potential targets for psychiatric drugs. Another, called corticotropin-releasing hormone (CRH) or corticotropin-releasing factor (CRF), was first identified in the hypothalamus, where it functions as part of the system that controls the pituitary gland. CRH signals the pituitary to release a peptide called ACTH, which in turn signals the outer layer (cortex) of the adrenal gland to secrete steroid hormones such as cortisol (hydrocortisone)—the same general plan that controls thyroxine secretion by way of TRH and TSH. As with the feedback control of TRH and TSH by thyroxine, there is also feedback control of CRH and ACTH by cortisol. But unlike the TRH-TSH system, which tends to keep a person's thyroxine blood levels within a fairly narrow range, the CRH-ACTH system is designed to induce massive secretion of cortisol into the blood in response to physical or psychological stress. The increased cortisol then makes its way to various organs, including the brain, and induces adaptive responses.

Psychiatrists have known for many years that about half of their severely depressed patients have a high blood level of cortisol.[†] They have also known that a potent synthetic cortisol-like hormone, called dexamethasone, which stops the secretion of cortisol in normal people by feedback to the CRH-ACTH system, doesn't stop the secretion of cortisol in many of the depressed patients whose cortisol levels are elevated. This suggests that these patients have an abnormality in their hypothalamus that makes it unresponsive to cortisol feedback. It also raised the possibility that the excess of cortisol in their blood may contribute to the pathophysiology of depression.[†] If this is the case, drugs that block CRH stimulation of the pituitary, and reduce blood cortisol, might alleviate depression.

Another reason that psychiatrists have become interested in CRH is that it does more than control the secretion of a hormone. Unlike TRH, which is confined to the hypothalamus, CRH is found in brain regions that participate in circuits that control emotions. These circuits also contain two receptors for CRH, called the CRH-1 and CRH-2 receptors. Both of these receptors have been implicated in emotional reactions[†] because knockout mice that lack either of them have altered behavioral responses to stressful situations. A role for the CRH system in human emotions is indicated by the finding that some depressed patients secrete excessive amounts of CRH, and that this abnormality is corrected by treatment with existing antidepressants.

Persuaded that CRH plays a role in the control of mood, several pharmaceutical companies are now trying to develop drugs that bind to CRH receptors. This is an approach that I personally championed more than ten years ago while serving on the scientific advisory board of DuPont-Merck, a joint venture of those two giant companies. In the early 1990s our board asked the company's pharmaceutical chemists to create potent blockers of CRH receptors. Gratified by the chemists' success, we had high hopes that one of these chemicals would become a useful psychiatric drug. But despite these achievements and the subsequent creation of more chemicals of this type[†] by several companies, these hopes have not yet been fulfilled.

After more than a decade of research, the only evidence that such drugs may help patients with major depression comes from a study with R121919, a drug made by Neurocrine Biosciences that blocks the CRH-1 receptor. Conducted at the Max Planck Institute of Psychiatry in Munich, the study demonstrated that R121919 produced a substantial reduction in symptoms of both depression and anxiety[†] in these patients. Unfortunately the results are hard to interpret because this was an "open-label" study in which placebos were not used, and both the patients and the doctors knew that they were participating in an initial test of an experimental drug. Because a few drug-treated patients had some transient impairment of liver function, less toxic alternatives are presently being developed. No placebo-controlled double-blind studies of this or other CRH receptor inhibitors have yet been reported.

::

In addition to targeting new receptors, pharmaceutical companies are also looking for drugs that influence the chemical changes in nerve cells that result from neurotransmission. These chemical changes carry signals throughout the interior of nerve cells. Many of them involve cyclic AMP and cyclic GMP, the "second messengers" that I mentioned previously. Like the signals conveyed by neurotransmitters, those transmitted by cyclic AMP and cyclic GMP must eventually be terminated so that the nerve cell is freed to participate in a new signaling event. This is achieved by inactivating cyclic AMP or cyclic GMP with enzymes called phosphodiesterases. Controlling the actions of these enzymes with drugs should, therefore, influence the duration of signaling, which might have a therapeutic effect.

Strong support for this general idea came from an accidental discovery with UK-92-480, a phosphodiesterase inhibitor that Pfizer was testing as a potential treatment for coronary artery disease. Although the effects of this drug on the heart proved to be disappointing, it turned out to be remarkably useful for the treatment of a disorder that psychiatrists are frequently called upon to deal with. Called male erectile disorder, its essential feature in *DSM-IV* (diagnostic code 302.72) is "a persistent or recurrent inability to attain, or to maintain until the completion of the sexual activity, an adequate erection." Although it is often a manifestation of a general medical condition, such as vascular disease, erectile dysfunction may also have psychological origins. Whatever these origins, it can frequently be helped not only by psychological treatment but, as Pfizer was delighted to discover, by UK-92-480. Based on this accidental discovery, UK-92-480 became a blockbuster drug with the chemical name of sildenafil and the brand name Viagra.[†]

We now know that penile erection depends on relaxation of smooth muscle cells in the blood vessels of the penis, which increases blood flow to that organ. We also know that the chemical that maintains the relaxation of penile smooth muscle cells is cyclic GMP, and that the penile enzyme that terminates the action of the second messenger is an enzyme called type 5-cyclic GMP-specific phosphodiesterase.[†] It was Pfizer's good fortune that Viagra just happens to be a potent and specific inhibitor of this enzyme. By retarding the cleavage of penile cyclic GMP, the drug maintains penile erection. Although this same enzyme is also present in other organs, they mainly rely on other phosphodiesterases to terminate the action of cyclic GMP. Therefore these other organs are not greatly affected by the drug.

The success of Viagra rejuvenated earlier studies of the effects of phosphodiesterase inhibitors on neurotransmission in the brain. They had been initiated in the 1980s with the discovery that a drug called rolipram, an inhibitor of type 4-cyclic AMP-specific phosphodiesterase, has antidepressant effects.[†] This finding was not pursued at the time because rolipram has very unpleasant side effects, especially nausea. Now the search for phosphodiesterase inhibitors for depression has intensified.[†]

It will not, however, have escaped your attention that the lucky discovery of Viagra in the mid-1990s, and the research on rolipram-like drugs that it has stimulated, is reminiscent of the lucky discovery of chlorpromazine and the fruitful research that it stimulated more than forty years earlier. Nowadays it is, of course, easier to build upon an accidental breakthrough because of the great advances that have been made in biology and chemistry. But the story of Viagra serves as a reminder that until there is a better understanding of the origins and pathophysiology of mental disorders, we will continue to depend on strokes of good luck to find important new psychiatric drugs.

11 :: *Clara's Prospects*

> The men of experiment are like the ant, they only collect and use; the
> reasoners resemble spiders, who make cobwebs out of their own
> substance. But the bee takes the middle course: it gathers its material
> from the flowers of the garden and field, but transforms and digests it by
> a power of its own. Not unlike this is the true business of philosophy
> [science]; for it neither relies solely or chiefly on the powers of the mind,
> nor does it take the matter which it gathers from natural history and
> mechanical experiments and lay up in the memory whole, as it finds it,
> but lays it up in the understanding altered and digested. Therefore, from
> a closer and purer league between these two faculties, the experimental
> and the rational, much may be hoped.
>
> —Francis Bacon (1620)

I opened this book with a description of Clara's mental symptoms
and their dramatic relief by manipulation of her neurotransmitters
with Prozac. I went on to describe the discovery of this and other
widely used psychiatric drugs, and the ways that new ones are being
developed. Now I turn to the prospects of fulfilling Clara's long-
standing hope for a treatment that is better than Prozac.

When Clara first expressed that hope, in the summer of 1994, I
advised her to be patient, because too little was known about the
reasons for her preoccupation with her nose to guide a search for
truly novel medications. In the years since then scientists have discovered some clues about Clara's disorder, as I will soon explain.

Nevertheless, translation of these discoveries into effective new remedies will still take a long time.

Meanwhile, the manipulation of Clara's neurotransmitters continues. Because she has been so troubled by Prozac-induced sexual dysfunction, I asked her to consult a psychiatrist who specializes in the treatment of female sexual problems. The consultant suggested that Clara add a small dose of bupropion to her daily Prozac. This unusual drug is approved for use both as an antidepressant (with the proprietary name of Wellbutrin) and to help people stop smoking by reducing craving for cigarettes (with the proprietary name of Zyban). In addition, as the consultant reminded me, bupropion is frequently prescribed off-label as a remedy for the sexual side effects of SSRIs, even though this use of the drug has not been tested in placebo-controlled trials. When I discussed this supplementary medication with Clara she was eager to try it.

Unfortunately, bupropion didn't relieve Clara's sexual symptoms. But, as with her initial experience with Prozac, there was a surprise: bupropion diminished the "dull-witted indifference" that she has complained of since taking the SSRI. This welcome result may be due to bupropion's augmentation of neurotransmission by dopamine and norepinephrine, both of which activate brain circuits that control attention and motivation. Although the small dose of bupropion may really be no more than a placebo, Clara is so pleased with it that I have continued her prescription for this additional medication.

Clara's use of a combination of psychiatric drugs is not unusual. In a 1997 study of the practice of 417 psychiatrists[†] sponsored by the American Psychiatric Association, more than half of their patients were receiving a combination of two psychiatric drugs, and almost a third were receiving three. Yet there have been very few formal clinical trials that have established the effectiveness of combinations of these medications,[†] and only limited studies of their interactions. More than half a century after the first psychiatric drugs were accidentally discovered, most patients continue to be treated by doctors whose prescriptions are guided by personal experience and the clinical judgment of experts, rather than on the basis of an understanding of the pathophysiology of mental disorders.

::

It is hard to predict how much longer it will take to achieve the requisite understanding. As part of this process, psychiatrists are rethinking the criteria they use to classify mental disorders. In Clara's case, her preoccupation with her nose, and her compulsive trips to the mirror, are defining features of the entity presently called body dysmorphic disorder (BDD). Yet Clara has a great deal in common with patients whose obsessive preoccupations and compulsive behaviors don't revolve around personal appearance, and who receive the diagnosis of obsessive-compulsive disorder (OCD). Because of these similarities, Eric Hollander of Mount Sinai School of Medicine and Katherine Phillips of Brown University have suggested that BDD and OCD might be combined into an overarching category called obsessive-compulsive (OC) spectrum disorders.[†] The advantage to Clara of being reclassified is that it removes her from the little-studied category called BDD and makes her eligible to benefit from the extensive ongoing research on OCD, which affects about 2 percent of people. As more information accumulates about the origins of BDD and OCD, the decision can be made about whether or not to keep them together.

Some of this information is coming from imaging studies of the brains of patients with OCD. For example, Lewis Baxter and his colleagues have discovered that these patients have unusually high nerve cell activity in brain regions[†] called the orbito-frontal cortex and the caudate nucleus, which are part of a brain circuit sometimes called the "OCD circuit."[†] This circuit has also been implicated in the distinctive symptoms of Tourette's disorder, which affects about one person in two thousand. Instead of the uncontrollable thoughts and rituals of OCD, people with Tourette's disorder have uncontrollable vocalizations, such as the uttering of grunts or obscenities, as well as other abnormal movements called tics.

The relationship between OCD and Tourette's disorder is more than anatomical and physiological. Evidence that these disorders may also have a shared genetic predisposition[†] comes from studies of their distribution in families. For example, David Pauls, James Leckman, and colleagues at Yale University found that families with members who have one of these disorders are more likely to also have members with the other disorder. Furthermore, some members of these families have symptoms of both disorders. Recognizing these relationships has led to the tentative inclusion of Tourette's disorder in the category of obsessive-compulsive spectrum disorders.

Lumping together such seemingly different disorders in a single category on the basis of anatomical, physiological, and genetic relationships is a radical departure from the way that psychiatric disorders are presently classified. In *DSM-IV*, diagnostic categories are based on patterns of symptoms, and the uncontrollable tics and utterances of Tourette's disorder are easy to separate from uncontrollable obsessions and compulsions. But the evidence that the same gene variants may change the "OCD circuit" in ways that result in these different types of episodic loss of control is opening the way to the discovery of the origins of these symptoms, and to new forms of treatment.

::

The search for gene variants that increase the risk of developing OCD and Tourette's disorder, which is just getting under way, recently got a big boost from an unexpected source: animal psychiatry. Working at the University of Utah, Mario Capecchi, an inventor of the technique for knocking out mouse genes, and his trainee, Joy Greer, found a remarkable behavioral effect of knocking out a gene called Hoxb8—a member of a group of genes that control the positioning of cells in the body. Unlike mice who lack other Hox genes, those lacking Hoxb8 did not have any observable malformations of body parts. Instead the only abnormality that Greer and Capecchi noticed was that the Hoxb8-knockout mice had some prominent bald patches.[†]

To discover the origin of these bald patches, Greer and Capecchi videotaped the mice for twenty-four hours—the same procedure that was used to detect narcolepsy in the hypocretin knockout mice. They were surprised to find that the Hoxb8 knockout mice had a distinctive form of obsessive-compulsive behavior: they kept vigorously grooming themselves and pulling out their hair! Furthermore, when

Greer and Capecchi housed a Hoxb8-knockout mouse with a normal mouse, the knockout mouse spent more time grooming the normal cagemate. This implies that the knockout mouse has a greater propensity to engage in grooming behavior and is not simply pulling out its hair because of itchy skin.

Based on these results, Greer and Capecchi concluded that the Hoxb8 gene must somehow participate in the establishment and maintenance of the nerve cell network that controls mouse grooming—an innate and stereotyped pattern of physical movements. Just as other Hox genes control cellular connections that determine the shape of body parts, Hoxb8 apparently controls connections of nerve cells in a circuit that is in charge of a pattern of behavior. In support of this idea, Greer and Capecchi demonstrated that the Hoxb8 gene is prominently expressed in nerve cells in the "OCD circuit" of normal adult mice, implying that it plays an important role in that circuit. They also point out that the bald spots they observed are reminiscent of those found in a human disorder called trichotillomania, a member of the OC spectrum disorders,[†] whose essential feature in *DSM-IV* is "the recurrent pulling out of one's own hair that results in noticeable hair loss."

The discovery that a single gene plays a critical role in controlling the intensity of a normal pattern of behavior—mouse grooming—has important implications for research on OC spectrum disorders. Now that the gene has been identified, scientists can study its effects on the establishment of connections in the OCD circuit and the way that it interacts with other genes that participate in this process. They can also find out if and how altered connections manifest themselves as a change in behavioral intensity.

This type of analysis may have important implications for other psychiatric disorders, which can also be fruitfully thought of as differences in the intensity of normal patterns of behavior—a dimensional instead of a categorical view of mental disorders. In this view major depression is an extreme version of the loss of hope that we all experience from time to time; social phobia is an extreme version of more common forms of shyness; even paranoid delusions can be considered on a continuum with normal suspiciousness. Therefore, understanding how a gene influences the intensity of a normal mouse behavior by controlling the connections of nerve cells in a brain circuit—and not just the actions of neurotransmitters—may shed light on many forms of psychopathology. Even though Hoxb8 is only one of a large number of genes that influence the development of brain circuits, the work with Hoxb8 may open up lines of research that change our ideas about the nature of mental illness.

In considering a more immediate clinical application of the Hoxb8 knockout mice, Greer and Capecchi suggest that they might be used to test the effects of drugs for the treatment of OC spectrum disorders. Presently these tests are all done with patient volunteers. As the brain molecules that are influenced by the Hoxb8 gene are identified, some may become targets for drug development. Having large numbers of knockout mice available for the screening of potential drugs will accelerate the discovery of new types of medications.

::

Despite these exciting speculations, Clara will still have to be patient. Finding potential drug targets through genetic research is only one step in the lengthy process of bringing an effective drug to market. As the ongoing work on secretase inhibitors for Alzheimer's disease illustrates, it takes many years to create chemicals that interact with newly identified targets, and to find out if they are clinically useful.[†] The difficulty of translating discoveries about the origins or pathophysiology of a disorder into practical remedies was recently emphasized by the editors of *Nature* as they launched a new periodical, called *Nature Reviews Drug Discovery*, that is devoted to the creation of drugs:

> There is little doubt that the drug-discovery business faces trouble ahead.[†] Despite huge increases in research spending, the pharmaceutical industry produces roughly the same number of new drugs each year, and there is little evidence that things are about to get better. . . . Although thousands of novel targets will be revealed over the coming years, a lead com-

pound [i.e., an initial drug candidate] with affinity for an isolated protein is a very long way from a drug.

Meanwhile, the search continues for ways to improve current drug treatments for patients with obsessions and compulsions.[†] For example, studies are under way with drugs that directly affect only one or a few of the fourteen or more serotonin receptors, in the hope of increasing therapeutic responses and reducing side effects. Combinations of SSRIs with benzodiazepines or atypical antipsychotics are also being studied.

So too are combinations of drugs with a specialized form of cognitive-behavior psychotherapy[†] called exposure and response prevention (ERP). Its aim is to teach OCD patients to refrain from compulsive behavior and to reduce the distress provoked by their obsessive thoughts. In the hands of experts who specialize in ERP it can be as effective as medications. But about a third of patients refuse this form of treatment, and many more drop out. Ongoing studies are assessing the effect of combining this form of psychotherapy with drugs such as clomipramine or Prozac.

A discouraging finding of all these studies is that patients who are greatly benefited by drugs or psychotherapy are rarely completely freed of symptoms. Even Clara still worries a little about her nose. But despite this lingering distress and the side effects of her medications, she is well aware that she is one of the very lucky ones who have responded well to treatment.

::

Among the great pleasures of meeting with Clara is that we sometimes talk about the philosophy of science, which remains her main professional interest. It was she who pointed out to me the quote from Francis Bacon that opens this chapter, about ants as experimentalists who only collect data, spiders as spinners of theories "out of their own substance," and bees as synthesizers "who transform and digest." Clara and I sometimes joke about the applicability of Bacon's vivid (albeit entomologically naive) metaphor to the activities of the people who are currently engaged in psychiatric research.

All too many of us psychiatric researchers usually conduct our business like Bacon's ants. We collect bits of information about mental symptoms and organize them into the patterns that delimit contemporary descriptions of mental disorders. We also test new remedies by trial and error, shrugging off our ignorance of the origins of the disorders we seek to treat.

Other psychiatric researchers resemble Bacon's spiders. They spin theoretical webs that frequently fail to stand up to rigorous critical scrutiny. Some of the spiders were educated in the academies of spinning that were founded by the greatest psychiatric spider, Sigmund Freud. But a growing number of psychiatric spiders have learned to spin with biological instead of psychological threads, replacing terms such as "Oedipus complex" with others such as "chemical imbalance."

All of us would like to behave more like Bacon's bees. One of the most notable successes of the psychiatric bees was the discovery that neurotransmission is influenced by the drugs, such as chlorpromazine and imipramine, whose therapeutic effects had been discovered by accident in the 1950s. Although the discoveries about neurotransmission didn't elucidate the origins of psychiatric disorders, they provided a rationale for the treatment of their main symptoms, and led to the creation of other valuable medications. But further bee work was constrained by the complexity of psychiatric problems and the lack of adequate tools with which to solve them.

Now the prospects for the bees are improving. The recent development of techniques for studying molecules, cells, circuits, and their roles in the operation of living human brains, is attracting a swarm of new bees. Many—like Greer and Capecchi—were trained in seemingly distant fields, such as genetics and developmental biology. Although they are not psychiatrists or psychologists, they are applying their special skills to the study of mental disorders. Some of them are beginning to analyze the complex sequences of genetically programmed and environmentally influenced brain processes that begin in the womb and go on through adolescence, when so many mental symptoms become apparent. In the course of this analysis they will identify the origins of the individual variations in brain functions that give rise to abnormal patterns of behavior. This, in turn, may finally free psychopharmacology from its fixation on neurotransmission, and open up new ways of treating patients such as Clara.

The results will not come all at once. It may take many years to accumulate and integrate the relevant material. It may take many more years to translate these discoveries into new psychiatric remedies. But the bees are hard at work. While we wait for them to succeed, Clara remains grateful that she can continue to rely on the benefits of her crude and imperfect medications, and on the considerable achievements of the ants and the spiders.

Appendix

Trade Names of Psychiatric Drugs

Generic Name	United States	United Kingdom
Antipsychotics (Typical)		
Chlorpromazine	Thorazine	Largactil
Fluphenazine	Prolixin	Moditen, Modecate
Haloperidol	Haldol	Haldol, Serenace
Antipsychotics (Atypical)		
Clozapine	Clozaril	Clozaril
Olanzapine	Zyprexa	Zyprexa
Quetiapine	Seroquel	Seroquel
Risperidone	Risperdal	Risperdal
Ziprasidone	Geodon	Zeldox
Tricyclic Antidepressants (TCAs)		
Amitriptyline	Elavil, Endep	Tryptizol, Lentizol
Clomipramine	Anafranil	Anafranil
Desipramine	Norpramin, Pertrofran	Norpramin, Pertrofran
Imipramine	Tofranil	Tofranil
Nortriptyline	Aventyl, Pamelor	Allegron
Monoamine Oxidase Inhibitors (MAOIs)		
Tranylcypromine	Parnate	Parnate

Generic Name	United States	United Kingdom
Selective Serotonin Reuptake Inhibitors (SSRIs)		
Citalopram	Celexa	Cipramil
Fluoxetine	Prozac, Sarafem	Prozac
Fluvoxamine	Luvox	Faverin
Paroxetine	Paxil	Seroxat
Sertraline	Zoloft	Lustral
Other Antidepressants		
Bupropion	Wellbutrin	
Mirtazapine	Remeron	Zispin
Nefazodone	Serzone	Dutonin
Venlafaxine	Effexor	Efexor
Mood Stabilizers		
Carbamazepine	Tegretol	Tegretol
Lithium carbonate	Eskalith, Lithobid	Camcolith, Priadel
Valproic acid	Depakote	Epilim
Benzodiazepines		
Alprazolam	Xanax	Xanax
Chlordiazepoxide	Librium	Librium
Clonazepam	Klonopin	Klonopin
Diazepam	Valium	Valium
Stimulants		
Amphetamine	Dexedrine, Adderall	Dexedrine
Methylphenidate	Ritalin, Concerta, Metadate	Ritalin, Equasym

Notes

Prologue

p. xiii. identified in the early 1970s. Wong et al. (1995).

p. xiii. transformed the practice of psychiatry. Snyder (1986), Kramer (1993), Barondes (1993, 1994).

p. xiv. the evolution of those that are widely prescribed today. Details about the history of these drugs are in Leake (1958), Swazey (1974), Smith (1991), Shorter (1997), and Healy (1997, 2002).

p. xiv. As you grow familiar with these fascinating chemicals. Further information about psychiatric drugs is in textbooks by Stahl (2000) and Nestler et al. (2001). Recent research is summarized in 134 chapters and 2010 pages by Davis et al. (2002).

Chapter 1

p. 4. a technique called cognitive therapy. Cognitive therapy, which is widely used to treat depression, is described by Beck and Beck (1995), Burns (1999), and in an overview of different forms of psychotherapy by Frank and Frank (1991).

p. 7. an off-label (i.e., not officially approved) use. In the United States each prescription drug is officially approved for the treatment of one or more carefully specified disorders by the Food and Drug Administration (FDA). Nevertheless, its prescription for other purposes is permitted and is very common. Such off-label use does not carry FDA approval and cannot be mentioned in the drug's package insert.

p. 12. many people who suffer from major depression start feeling better while taking a placebo. Brown (1998), Quitkin et al. (2000).

p. 12. it is not clear which individuals get better because of a placebo effect and which get better because of a drug effect. One way to tell them apart is that placebo effects are more likely to peter out with prolonged treatment (Quitkin et al. [1993]).

p. 14. It is well known that antidepressants usually don't have much effect until they have been taken for several weeks, whereas placebo effects tend to come on more quickly. Quitkin et al. (1984).

p. 15. Her main diagnosis wasn't dysthymic disorder. In psychiatry, as in other branches of medicine, a patient may be given more than one diagnosis (Kessler et al. [1994], Kaufman and Charney [2000]).

p. 15. a variant called body dysmorphic disorder. Phillips (1986). I will return to the similarities between body dysmorphic disorder and obsessive-compulsive disorder in Chapter 11.

p. 15. This was subsequently confirmed by a series of double-blind placebo-controlled studies . . . with Prozac. Phillips et al. (2002).

Chapter 2

p. 17. Numerous devious pathways. The quote is from Kety's foreword to Swazey (1974).

p. 17. Introduced in 1912 by Bayer. Shorter (1997).

p. 18. The discovery of this remarkable medication. Swazey (1974), Healy (2002).

p. 18. On October 3, 1950, he circulated a memo. The text of the memo is in Swazey (1974).

p. 19. Among the psychiatric pioneers was Pierre Deniker. Delay, Deniker, and Harl (1952). Deniker's reminiscences about early studies with chlorpromazine are in Ayd and Blackwell (1970).

p. 20. By 1955 it was being administered to hundreds of thousands of people with schizophrenia. Despite its widespread use, the effectiveness of chlorpromazine was not firmly established until publication of the findings of the National Institute of Mental Health Psychopharmacology Service Center Collaborative Study Group (1964).

p. 22. Much of the credit for solving this mystery in the 1950s and 1960s goes to Arvid Carlsson. Snyder (1986).

p. 22. Neurotransmitters are the chemicals that nerve cells use to communicate with each other. Snyder (1986), Barondes (1993), Nestler et al. (2001), Davis et al. (2002).

p. 24. methods for measuring binding of dopamine to receptors. Creese et al. (1976), Seeman et al. (1976).

p. 24. the dopamine hypothesis of schizophrenia. Meltzer and Stahl (1976). This hypothesis, which owes so much to the work of Arvid Carlsson, is now considered simplistic. Carlsson's recent view of neurotransmission in schizophrenia is summarized in Carlsson et al. (2000).

p. 25. whereas their reduction of delusions and hallucinations develops over many weeks. Lieberman et al. (1993), Garver et al. (1997).

p. 25. When this was shown to be the case in 1988. Kane et al. (1988).

p. 28. that have many of the favorable features of clozapine. Some patients who are helped by clozapine get equivalent relief from other atypical antipsychotics (Rafal et al. [1999]). But some are only helped by clozapine (Conley et al. [1999]). The special therapeutic properties of clozapine may be due, in part, to its effects on receptors other than those for dopamine and serotonin (Litman et al. [1996]).

p. 29. partial dopamine agonists. Tamminga (2002).

p. 29. dopamine system stabilizers. Stahl (2001).

p. 29. aripiprazole . . . was developed in Japan. Kikuchi et al. (1995).

p. 29. the NMDA receptor. The reasons that this receptor may be a good target for antipsychotic drugs are summarized by Heresco-Levy (2000) and by Snyder and Ferris (2000).

p. 29. Among phencyclidine's effects are hallucinations and a sense of detachment. Heresco-Levy (2000), Snyder and Ferris (2000).

p. 30. a specialized part of the NMDA receptor that binds glycine or D-serine. Krystal and D'Souza (1998), Heresco-Levy (2000), Snyder and Ferris (2000).

p. 30. D-serine produced some additional improvement. Tsai et al. (1998).

Chapter 3

p. 32. As Kuhn later recalled. Kuhn, in Ayd and Blackwell (1970).

p. 33. description of his results. Kuhn (1958).

p. 35. iproniazid . . . used widely. Kline, in Ayd and Blackwell (1970).

p. 36. The godfather of this stimulating milieu was Bernard Brodie. Kanigel (1986) describes Brodie's role in the creation of molecular psychopharmacology.

p. 36. Axelrod made an even more remarkable discovery. Axelrod (1961).

p. 38. the norepinephrine hypothesis of depression. Bunney and Davis (1965), Schildkraut (1965), Schildkraut and Kety (1967).

p. 38. imipramine blocks reuptake of serotonin. Axelrod and Inscoe (1963).

p. 40. zimelidine—the first selective serotonin reuptake inhibitor (SSRI)—relieved depression. The story of zimelidine is in Healy (1997).

p. 40. Lilly 110140 is a potent inhibitor of serotonin reuptake. Wong et al. (1995).

p. 42. a rating scale for depression. Hamilton (1960).

p. 42. confirmed in studies of severely depressed patients. Placebo-controlled studies with imipramine are summarized by Quitkin et al. (2000).

p. 43. some depressed patients who received sugar pills for two months were also feeling a lot better. The fact that depressed patients often get better while taking a placebo complicates studies of the therapeutic effects of antidepressants. Quitkin et al. (2000) consider some of the problems in interpreting the results of placebo-controlled trials of these drugs.

p. 43. or even surgical placebo. For example, a recent study (Moseley et al. [2002]) showed that superficial surgical incisions around the knee, a placebo treatment, are as effective in relieving symptoms of arthritis as authentic arthroscopic surgery of the knee joint, a widely used treatment. Brown (1998) reviews evidence that "the placebo effect is not unique to psychiatric illness. For a wide range of afflictions, 30 to 40 percent of patients experience relief after taking a placebo."

p. 43. which have a variety of effects on neurotransmission. The many effects of these drugs are reviewed by Stahl (2000).

p. 43. Some are helped by another antidepressant. Joyce et al. (1994) present evidence that depressed patients with certain temperaments (as defined with elaborate rating scales) respond better to a drug that augments neurotransmission by norepinephrine, whereas those with other temperaments respond better to drugs that augment neurotransmission by serotonin. Akiskal (1995) suggests that depression can be divided into subtypes based on the temperament of the patient and response to particular antidepressant medications.

p. 44. patients with dysthymic disorder . . . are indeed helped by antidepressants. Thase et al. (1996). But the difference between antidepressants and placebos is greater in studies of major depression (Khan et al. [2002]).

p. 45. The biggest mystery about antidepressants is that it generally takes weeks or more of continuous drug treatment for the therapeutic effect to develop. Duman et al. (1997).

p. 45. It is this new state of cellular chemistry that is believed to be responsible for the antidepressant effect. Duman et al. (1997), Manji et al. (2001).

p. 46. a protein called BDNF. Duman et al. (1997), Duman (1998).

p. 46. The same is true of antipsychotic drugs. The delayed therapeutic effects of antipsychotic drugs, which were first observed in the 1950s, have been documented in many studies such as those by Lieberman et al. (1993) and Garver et al. (1997).

p. 46. It is also true of the drugs that are used to treat a distinctive mood disorder called bipolar disorder. Manji and Zarate (2002).

p. 46. Discovered by the same sorts of accidents. The discovery that lithium (Cade 1949), valproic acid (Lambert et al. [1966]), and carbamazepine (Okuma et al. [1973]), relieve mania were all based on accidental observations. The current use of these and related drugs such as gabapentin (Neurontin), lamotrigine (Lamictal), and topiramate (Topamax) for the treatment of bipolar disorder is reviewed by Brambilla et al. (2001).

p. 46. they work on the internal nerve cell machinery. Manji and Zarate (2002).

Chapter 4

p. 47. "It would be wrong and naive." Berger, in Ayd and Blackwell (1970).

p. 49. Newer psychotherapies for anxiety, such as cognitive therapy and behavioral therapy. Frank and Frank (1991).

p. 50. B. J. Ludwig, created the derivative that became known as meprobamate. Berger, in Ayd and Blackwell (1970), Smith (1991).

p. 50. "Miltown was of considerable value." Selling (1955).

p. 51. In a paper describing the development of Miltown. Berger, in Ayd and Blackwell (1970).

p. 51. Miltown wasn't really much better than phenobarbital after all. Greenblatt and Shader (1971).

p. 52. The product of tinkering by Leo Sternbach. The discovery of benzodiazepines is described by Irvin M. Cohen in Ayd and Blackwell (1970), by Sternbach (1979), and by Smith (1991).

p. 54. Both Costa and Haefely showed that the benzodiazepines work by augmenting the actions of a neurotransmitter called gamma-aminobutyric acid. Costa et al. (1975), Haefely et al. (1975).

p. 55. Phenobarbital, meprobamate, and alcohol also change the shapes of certain GABA receptors. Rho et al. (1997).

p. 55. none of these drugs binds to the benzodiazepine binding site. The existence of this binding site on many GABA receptors raised the possibility that there is a natural benzodiazepine-like chemical in the brain that binds to this site and controls anxiety. But so far none has been found. There is, however, a hormone in the brain called allopregnanolone that augments GABA transmission by binding to GABA receptors in a different way than benzodiazepines. Prozac increases the concentration of this hormone in the brain (Uzunova et al. [1998], Guidotti and Costa [1998]) and increases its manufacture (Griffin and Mellon [1999]), which may account for some of the drug's effects on anxiety.

p. 55. Called tolerance, this loss of effectiveness reflects counter-measures that the brain mounts to overcome the effects of the drug. Hyman and Nestler (1996), Stahl (2000), Nestler et al. (2001).

p. 56. In October 1958 they began a radical experiment. Klein and Fink (1962).

p. 57. So why call imipramine and MAO inhibitors antidepressants? Psychiatric drugs are generally classified into categories such as antipsychotics and antidepressants on the basis of their initial use or their commonest use. But, in practice, all psychiatric drugs are used to treat a variety of symptoms.

p. 58. the SSRIs are considered to be interchangeable. Although all of the SSRIs can be used to treat anxiety disorders, some psychiatrists tend to avoid using sertraline or Prozac for this purpose because they may have more activating effects during the first few weeks of treatment than paroxetine, fluvoxamine, or citalopram. With prolonged treatment all the SSRIs are believed to be equally effective, although patients may vary in the way they respond to a particular member of this class of drugs.

p. 58. whose symptoms justify the diagnosis of both a mood disorder and an anxiety disorder. The coexistence of anxiety and depression is very common (Kessler et al. [1994], Kaufman and Charney [2000]).

Chapter 5

p. 61. "the psychological response of children to drug therapy." Bradley (1937).

p. 62. "Possibly the most striking change in behavior." Bradley (1937).

p. 62. Gordon Alles, the chemist who had synthesized it. Leake (1958).

p. 63. "marks one of the triumphs attained in the field of synthetic chemistry." Chen and Kao (1926).

p. 64. use it for the treatment of narcolepsy. Prinzmetal and Bloomberg (1935).

p. 65. "seems to have a definite though limited value in combating the neuroses." Myerson (1936).

p. 66. "By a subdued response is meant." Bradley and Bowen (1941).

p. 68. increased sensitivity to certain actions of the drug, called sensitization. Unlike sedating drugs such as phenobarbital or Valium, which call forth brain countermeasures that *reduce* subsequent responsiveness to these drugs (called tolerance), stimulants such as amphetamine may elicit an *increase* in the responsiveness of the brain to the next dose. This process, called sensitization, plays a role in addiction to stimulants (Nestler et al. [2001]).

p. 68. gradual modifications in brain chemistry and physiology may lead to their compulsive use—the well-known pattern of behavior called addiction. Hyman and Nestler (1996), Nestler et al. (2001).

p. 70. In Bradley's time children with extreme forms of this pattern of behavior were said to have "minimal brain dysfunction." Although this diagnosis existed in Bradley's time, the disorder was rarely severe enough to lead to hospitalization. It is, therefore, very likely that the patients in his hospital who were the subjects of his experiments with amphetamine suffered from other behavioral disturbances. In retrospect, the favorable effects of amphetamine that he observed in his patients would not have suggested that this drug is a specific treatment for the children who would now be diagnosed as having ADHD.

p. 71. Although the current diagnostic scheme is categorical . . . many psychiatrists actually use a graded or dimensional diagnostic scheme for certain disorders. As explained in the introduction to the latest edition of American psychiatry's diagnostic manual, *DSM-IV* (American Psychiatric Association, 1994): "*DSM-IV* is a categorical classification that divides mental disorders into types based on criteria sets with defining features. This naming of categories is the traditional method of organizing and transmitting information in everyday life and has been the fundamental approach used in all systems of medical diagnosis. . . . It was suggested that the *DSM-IV* Classification be organized following a dimensional model rather than the categorical model. . . . A dimensional system classifies clinical presentations based on quantification of attributes rather than the assignment to categories and works best in describing phenomena that are distributed continuously and that do not have clear boundaries. Although dimensional systems increase reliability and communicate more clinical information (because they report clinical attributes that might be subthreshold in a categorical system), they also have serious limitations and thus far have been less useful than categorical systems in clinical practice and in stimulating research. Numerical dimensional descriptions are much less familiar and vivid than are the categorical names for mental disorders. Moreover there is as yet no agreement on the choice of the optimal dimensions to be used for classification purposes." For a discussion of the relative merits of dimensional and categorical classifications see Goldberg (2000).

p. 71. may also be correlated with desirable attributes. Jensen et al. (1997).

p. 72. A recent study of about six hundred children. MTA Cooperative Group (1999).

p. 72. One of them, Peter Jensen, summed it up this way. Cited in Psychiatric News (January 21, 2000) 20.

p. 73. "our data do not strongly support the concept of a unique stimulant drug response." Rapoport et al. (1980).

p. 74. some children with ADHD remain inattentive and impulsive into adulthood. Faraone et al. (2000).

p. 75. guanfacine . . . is a useful treatment for ADHD. Scahill et al. (2001).

p. 75. So too is desipramine. Wilens et al. (1996). Eli Lilly has just introduced atomoxetine, which blocks reuptake of norepinephrine more selectively than desipramine, as a new treatment for ADHD. Approved by the FDA for use in children, it is being marketed under the trade name of Strattera.

Chapter 6

p. 79. "ideal psychotherapeutic drug." Hollister (1975) concluded his article with the following paragraph: "The ideal psychotherapeutic drug would (1) cure or alleviate the pathogenetic mechanisms of the symptom or disorder; (2) be rapidly effective; (3) benefit most or all patients for whom it is indicated; (4) be non-habituating and lack potential for creating dependence; (5) not allow tolerance to develop; (6) have minimum toxicity in the therapeutic range; (7) have a low incidence of secondary side effects; (8) not be lethal in overdoses; (9) be adaptable both to inpatients and outpatients; (10) not impair any cognitive, perceptual, or motor functions. No such drug exists, but to a fairly surprising degree many of the available drugs meet the majority of these desiderata. It has been both our blessing and our curse that we had effective drug therapy for emotional disorders before we had a science of behavioral pathology. Our best hope for getting better psychotherapeutic drugs is to understand better the causes of emotional disorders."

p. 83. drugs that block the manufacture of thyroxine. Astwood (1945).

p. 86. the cause of the excessive thyroxine production could ultimately be attributed to sustained emotional distress. Alexander (1950).

p. 87. serum from Mrs. McCabe. Adams and Purves (1956).

p. 87. The identity of LATS, and the way that it affects the thyroid Adams (1988).

p. 87. LATS tricks the thyroid cell into responding as if it is being stimulated by TSH. In fact in the blood of patients with Graves' disease the level of authentic TSH is very low, because production of TSH is inhibited by the high levels of blood thyroxine, whose manufacture is stimulated by LATS.

p. 88. antibodies that bind to cells in the eye sockets. Bahn and Heufelder (1993).

p. 88. "Heredity, of course." Alexander (1950).

p. 89. which included 322 Hungarian women with Graves' disease. Stenszky et al. (1985).

p. 89. clustering in families could reflect shared environment as well as shared genes. Weetman (2001).

p. 89. some geneticists are so persuaded of the importance of genes. Brix et al. (2001).

p. 90. make this distinction between psychiatric disorders and medical disorders. In the past identification of a biological cause of a psychiatric disorder led to its reclassification as a medical disorder. A notable example is general paresis, which affected about 10 percent of patients in psychiatric hospitals in 1900. But the discovery that it is caused by infection of the brain (summarized in Barondes [1993]) immediately converted it to a medical disorder. This was partly due to the misconception that any disorder whose cause is biological can't be considered psychiatric.

p. 91. a recent study by a group of Robert Graves' Irish countrymen. Leary et al. (1999).

p. 92. X-ray treatments of the eye sockets. Gorman et al. (2001).

Chapter 7

p. 95. "At one end of the series." Freud (1917), Lecture 22. The full quote is: "From the point of view of causation, cases of neurotic illness fall into a *series* within which the two factors—sexual constitution [which is inherited] and events experienced, or, if you wish, fixation of libido and frustration—are represented in such a way that where one of them predominates the other is proportionately less pronounced. At one end of the series stand those extreme cases of whom one can say: These people would have fallen ill whatever happened, whatever they experienced, however merciful life has been to them, because of their anomalous libido development. At the other end stand cases which call forth the opposite verdict—they would undoubtedly have escaped illness if life had not put such and such burdens upon them."

p. 95. a middle-aged woman named Auguste D. The story of Auguste D. is told by Maurer et al. (1997) and by Tanzi and Parson (2000).

p. 97. A breakthrough came in the early 1980s. Glenner and Wong (1984).

p. 97. the DNA code for the amino acid valine (V) is CAA. There are actually alternative codes for some amino acids. For valine there are four alternatives, whereas for other amino acids such as methionine there is just one code.

p. 98. There are about thirty thousand human genes. The exact number of human genes is still not established. Current estimates range between thirty thousand and forty thousand.

p. 99. identified a variation in the gene for APP in the DNA from a British patient. Lendon et al. (1997).

p. 100. some other families had different changes in the gene for APP. Lendon et al. (1997), Tanzi and Parson (2000).

p. 100. presenilin-1 . . . presenilin-2. Lendon et al. (1997), Tanzi and Parson (2000).

p. 101. several other enzymes were discovered that also cut APP. Esler and Wolfe (2001).

p. 101. the neurotransmitter that is targeted is acetylcholine. Bartus et al. (1982).

p. 102. treatment with these drugs produces transient improvements. Rockwood et al. (2001), Rogers et al. (1998), Corey-Bloom et al. (1998).

p. 102. pharmaceutical companies are trying to make drugs that prevent the ongoing accumulation of amyloid. Selkoe (1999).

p. 103. the most important, called APOE. Strittmater and Roses (1996).

p. 104. people who take cholesterol-lowering drugs . . . have less Alzheimer's disease. Simons et al. (2001).

p. 104. an anti-inflammatory and analgesic drug that may impede the progression of Alzheimer's disease. Stewart et al. (1997).

p. 104. the increased concentration of certain mental disorders in families is impressive. Family studies of schizophrenia, depression, bipolar disorder, ADHD, and obsessive-compulsive disorder are summarized by Steven Moldin in Report of the National Institute of Mental Health's Genetics Workgroup (1999).

p. 105. two approaches are being used to find the gene variants that play a part in the development of mental disorders. For a description of the hunt for these gene variants, see Barondes (1998, 1999).

p. 106. people with schizophrenia are more likely to have a particular form of the enzyme. Egan et al. (2001).

p. 106. studying the combined effects of these two gene variants on the risk of ADHD. The ADHD Molecular Genetics Network (2001).

p. 109. a variant of a gene called neuregulin-1. Stefansson et al. (2002).

p. 109. has implicated a variant of a gene called dysbindin. Straub et al. (2002).

p. 109. identify all the gene variants. Risch (2000).

p. 110. also in their relatives with functional abnormalities of the prefrontal cortex. Egan et al. (2001).

p. 110. Using a dimensional approach may be necessary to find the gene variants that influence susceptibility to this and other mental disorders. Brown et al. (1992), Raeymaekers and Van Broeckhoven (1998).

Chapter 8

p. 113. a dog version of obsessive-compulsive disorder (OCD). Overall (2000).

p. 113. the canine version responds to drugs. Wynchank and Berk (1998), Hawson et al. (1998).

p. 117. set out to breed lines of mice to find out how fearful or fearless they would become. DeFries et al. (1978).

p. 118. scientists have identified the chromosomal locations of the genes that influence these patterns of behavior. Flint et al. (1995), Turri et al. (2001).

p. 118. techniques have been developed to create animals that have new gene variants. Tecott and Wehner (2001).

p. 118. by deliberately creating random mutations in the DNA of thousands of mice. A notable example is the mutation in a mouse gene that is essential for circadian behavior, such as the daily cycle of sleep and wakefulness. This random mutation, which was created by treating mice with ENU, led to the identification of the relevant gene, named Clock (Viaterna et al. [1994]) and to the subsequent identification of the same and related genes in humans. This discovery is stimulating research on the control of human rhythms, many of which are abnormal in patients with mental disorders such as depression.

p. 119. each of which gives rise to a form of Alzheimer's disease in the mice that receive it. Van Leuven (2000).

p. 119. lines of transgenic mice with both an APP gene variant and a PS1 gene variant. Borchelt et al. (1997), Dewachter et al. (2001).

p. 121. they were amazed to find that the mice kept falling asleep for short periods. Chemelli et al. (1999).

p. 121. this variant encodes an inactive form of a receptor for one of the hypocretins. Lin et al. (1999).

p. 122. the brains of people with narcolepsy do indeed have abnormally small amounts of hypocretins. Siegel et al. (2001).

p. 122. the hypocretin deficiency in human narcolepsy is due to an autoimmune reaction. Lin et al. (2001).

p. 123. a relatively new drug for narcolepsy called modafinil (Provigil), discovered in the 1980s. Bastuji and Jouvet (1988).

p. 123. new techniques are available for reversibly knocking out genes and for selectively knocking them out in particular brain regions. Tecott and Wehner (2001), Gross et al. (2002).

p. 125. They replaced the normal gene for the alpha-1 protein with a variant that doesn't bind benzodiazepines. Rudolph et al. (1999).

p. 125. L838,417 does not produce sedation, because it doesn't bind the alpha-1 protein. McKernan et al. (2000).

p. 126. the anxiety-reducing effect depends on binding to alpha-2. Low et al. (2000).

p. 127. the mouse genome project. Boguski (2002).

p. 127. identification of gene variants that influence mouse fearfulness. Flint et al. (1995), Turri et al. (2001).

Chapter 9

p. 129. If it were not for the great variability. The statement by Osler, the great nineteenth-century physician, is cited by Roses (2000).

p. 129. a homicide investigation was begun. Stipp (2000).

p. 130. In 1953 Axelrod began studying the ways that the body inactivates amphetamine. The history of this discovery is described in Kanigel (1986).

p. 131. they are toxic natural chemicals found in terrestrial plants. Nebert (1997).

p. 131. About half a dozen enzymes play major roles in the metabolism of the commonly used psychiatric drugs. Fromm et al. (1997).

p. 132. St. John's wort . . . stimulates the body to make more of the CYP3A enzymes. Vogel (2001).

p. 132. The discovery that some people have defective drug-metabolizing enzymes. Mahgoub et al. (1977).

p. 133. At least 25 percent of Caucasians have a defect in one of their 2D6 genes. Sachse et al. (1997).

p. 133. 29 percent of Ethiopians have the 2D6 gene variant that makes them ultrarapid metabolizers. Aklillu et al. (1996).

p. 134. a large percentage of Japanese and other Asians have inherited gene variants that produce defective aldehyde dehydrogenase. Higuchi et al. (1996).

p. 135. a federal jury in Cheyenne, Wyoming, ordered Glaxo-SmithKline . . . to pay $6.4 million to relatives of Donald Schell. Thompson (2001).

p. 136. another famous lawsuit about psychiatric drugs . . . by Rafael Osheroff. Klerman (1990).

p. 137. some of these variants affect the functions of these proteins. Cravchik and Goldman (2000).

p. 137. This physiological difference appears to have behavioral consequences. Murphy et al. (2001).

p. 137. depressed people who have a short variant of the serotonin transporter gene. Zanardi et al. (2001).

p. 138. bipolar patients who have SSRI-induced mania. Mundo et al. (2001).

p. 138. a variant of the gene for the D3 dopamine receptor appears to be implicated in the development of tardive dyskinesia. Ozdemir et al. (2001).

p. 139. a foundation for the gradual replacement of our one-size-fits-all medications. Roses (2000).

Chapter 10

p. 142. WY-45030 blocks the reuptake of both serotonin and norepinephrine Haskins et al. (1985).

p. 142. certain depressed patients get more complete relief from Effexor than from SSRIs. Thase et al. (2001).

p. 143. Lilly is . . . presently pinning its hopes on duloxetine. Goldstein et al. (2002).

p. 143. Duloxetine has a relatively greater effect on norepinephrine than Effexor. Bymaster et al. (2001).

p. 144. each of these drugs has already gone through a time-consuming and expensive procedure of testing. Bringing a new drug to market usually takes more than ten years from the first hints of its potential benefits. The process begins with an initial evaluation of its safety in experimental animals and proceeds through three phases of testing in humans (Thase [1999]). In Phase I, which takes about a year, safety and side-effects are evaluated in volunteers. In Phase II, which takes several more years, the drug is tested in a few hundred patients with a particular disorder to get some idea of its potential usefulness. In Phase III, which takes about three or four years, effectiveness is tested in a placebo-controlled study of about two thousand patients with a particular disorder. The results are evaluated by a panel of experts who advise the FDA about approval. This final approval process may take a few years.

p. 144. the manufacturer may be willing to support double-blind placebo-controlled studies to test the drug's effectiveness for this purpose. To get an established drug approved for the treatment of a different disorder, it is necessary to go through Phase III testing in patients with that disorder, which may take several years and cost more than a hundred million dollars.

p. 144. "paroxetine is an effective treatment for patients with generalized social phobia." Stein et al. (1998).

p. 146. MK-869 was as good an antidepressant as paroxetine (Paxil). Kramer et al. (1998).

p. 147. Merck is continuing with this program of research. Rupniak and Kramer (1999), Langreth (2002). Merck is also hoping to bring MK-869 to market for a completely different indication: as a treatment for the nausea and vomiting that is a serious side effect of other medical treatments, such as cancer chemotherapy.

p. 148. severely depressed patients have a high blood level of cortisol. Arborelius et al. (1999). The depressed patients with elevated
levels of cortisol are less likely to respond to a placebo than those
with normal levels of cortisol (Brown [1998]).

p. 148. the excess of cortisol in their blood may contribute to the
pathophysiology of depression. This has raised the possibility that
depression may be treated by blocking the body's manufacture
of cortisol. (Reus and Wolkowitz [2001], Belanoff et al. [2001]).

p. 148. CRH-1 and CRH-2 receptors . . . have been implicated in
emotional reactions. Contarino et al. (1999), Bale et al. (2000).

p. 149. the subsequent creation of more chemicals of this type.
Hodge et al. (1999), McCarthy et al. (1999).

p. 149. R121919 produced a substantial reduction in symptoms of
both depression and anxiety. Zobel et al. (2000).

p. 150. Based on this accidental discovery UK-92-480 became a
blockbuster drug . . . Viagra. Boolell et al. (1996).

p. 150. an enzyme called type 5-cyclic GMP-specific phosphodiesterase. Another potent inhibitor of this enzyme has been discovered by Hosogai et al. (2001).

p. 150. rolipram, an inhibitor of type 4-cyclic AMP-specific phosphodiesterase, has antidepressant effects. Bobon et al. (1988).
The drug first aroused interest because of its antidepressant
effects, and was only later found to be an inhibitor of phosphodiesterase.

p. 150. the search for phosphodiesterase inhibitors for depression
has intensified. Fujimaki (2000), Dyke and Montana (2002).

Chapter 11

p. 154. In a 1997 study of the practice of 417 psychiatrists. Pincus et
al. (1999).

p. 154. there have been very few formal clinical trials that have established the effectiveness of combinations of these medications. A
rationale for combining psychiatric drugs is offered by Stahl
(2000), who considers many combinations, including some—
called "heroic combos"—that he suggests for patients who don't
get relief from a single medication.

p. 155. combined into an overarching category called obsessive-compulsive (OC) spectrum disorders. Phillips et al. (1995), Hollander and Wong (2000). In addition to the behavioral similarities of BDD and OCD, they share a distinctive response to drugs. Unlike depression, which responds to drugs that augment neurotransmission by either serotonin (such as Prozac) or norepinephrine (such as desipramine), both BDD and OCD respond only to drugs that augment neurotransmission by serotonin (Goodman et al. [1990], Hollander et al. [1999]). This suggests similarities in the pathophysiology of BDD and OCD.

p. 155. these patients have unusually high nerve cell activity in brain regions. Saxena et al. (1998), Baxter et al. (2000).

p. 155. the "OCD circuit." Graybiel and Rauch (2000), Graybiel and Saka (2002).

p. 155. OCD and Tourette's disorder . . . may also have a shared genetic predisposition. Pauls et al. (1995).

p. 156. the Hoxb8-knockout mice had some prominent bald patches. Greer and Capecchi (2002).

p. 157. trichotillomania, a member of the OC spectrum disorders. Hollander and Wong (2000).

p. 158. it takes many years to create chemicals that interact with newly identified targets, and to find out if they are clinically useful. Hardy and Selkoe (2002).

p. 158. There is little doubt that the drug-discovery business faces trouble ahead. Nature, 415 (January 3, 2002), 1.

p. 159. the search continues for ways to improve current drug treatments for patients with obsessions and compulsions. Blier et al. (2000).

p. 159. combinations of drugs with a specialized form of cognitive-behavior psychotherapy. Kozak et al. (2000).

Bibliography

Adams DD. (1988). Long-acting thyroid stimulator: how receptor autoimmunity was discovered. Autoimmunity, 1, 33–39.

Adams DD, Purves HD. (1956). Abnormal responses in the assay of thyrotrophin. Proceedings of University of Otago Medical School, 34, 11–12.

ADHD Molecular Genetics Network (2001). Report from the second international meeting of the attention deficit hyperactivity disorder molecular genetics network. American Journal of Medical Genetics (Neuropsychiatric Genetics), 105, 225–228.

Akiskal HS. (1995). Toward a temperament-based approach to depression: implications for neurobiological research. Adv Biochem Pharmacol, 49, 99–112.

Aklillu E, Persson I, Bertilsson L, Johansson I, Rodrigues F, Ingelman-Sundberg M. (1996). Frequent distribution of ultrarapid metabolizers of debrisoquine in an Ethiopian population carrying duplicated and multiduplicated functional CYP2D6 alleles. J Pharmacol Exp Ther, 278, 441–446.

Alexander F. (1950). Psychosomatic Medicine, Its Principles and Applications. New York, Norton.

American Psychiatric Association. (1994). Diagnostic and Statistical Manual of Mental Disorders (4th edition) (DSM-IV). Washington, D.C., American Psychiatric Press.

American Psychiatric Association. (1998). Practice guidelines for the treatment of panic disorder. Washington, D.C., American Psychiatric Press.

Arborelius L, Owens MJ, Plotsky PM, Nemeroff CB. (1999). The role of corticotropin-releasing factor in depression and anxiety disorders. J Endocrinol, 160, 1–12.

Astwood EB. (1945). Chemotherapy of hyperthyroidism. Harvey Lectures, 40, 195–235.

Axelrod J. (1961). Effect of psychotropic drugs on the uptake of tritiated noradrenaline by tissues. Science, 133, 383–384.

Axelrod J, Inscoe JK. (1963). The uptake and binding of circulating serotonin in the effect of drugs. J Pharmacol Exp Ther, 141, 161–165.

Ayd FJ, Blackwell B. (1970). *Discoveries in Biological Psychiatry.* Philadelphia, J.B. Lippincott.

Bacon F. (1620). *The New Organon, and Related Writings.* Edited by FH Anderson. New York, Liberal Arts Press (1960).

Bahn RS, Heufelder AE. (1993). Pathogenesis of Graves' ophthalmopathy. New Engl J Medicine, 329, 1468–1475.

Bale TL, Contarino A, Smith GW, Chan R, Gold LH, Sawchenko PE, Koob GF, Vale WW, Lee K-F. (2000). Mice deficient for corticotropin-releasing hormone receptor-2 display anxiety-like behaviour and are hypersensitive to stress. Nature Genetics, 24, 410–414.

Barondes SH. (1993). *Molecules and Mental Illness.* New York, Scientific American Library.

Barondes SH. (1994). Thinking about Prozac. Science, 263, 1102–1103.

Barondes SH. (1998). *Mood Genes: Hunting for Origins of Mania and Depression.* New York, WH Freeman.

Barondes SH. (1999). An agenda for psychiatric genetics. Arch Gen Psychiatry, 56, 549–552.

Bartus RT, Dean RL, Beer B, Lippa S. (1982). The cholinergic hypothesis of geriatric memory dysfunction. Science, 217, 408–414.

Bastuji H, Jouvet M. (1988). Successful treatment of idiopathic hypersomnia and narcolepsy with modafinil. Prog Neuropsychopharmacol Biol Psychiatry, 12, 695–700.

Baxter LR Jr, Ackermann RF, Swerdlow NR, Brody A, Saxena S, Schwartz JM, Gregoritch JM, Stoessel P, Phelps ME. (2000).

Specific brain system mediation of obsessive-compulsive disorder responsive to either medication or behavior therapy. In Goodman et al., 573–609.

Beck JS, Beck AT. (1995). *Cognitive Therapy*. New York, Guilford Press.

Belanoff JK, Flores BH, Kalezhan M, Sund B, Schatzberg AF. (2001). Rapid reversal of psychotic depression using mifepristone. J Clin Psychopharmacol, 21, 516–521.

Berger FM. (1970). Anxiety and the discovery of tranquilizers. In Ayd FJ, Blackwell B, 115–141.

Blier P, Bergeron R, Pineyro G, El Mansari M. (2000). Understanding the mechanism of action of serotonin reuptake inhibitors in obsessive-compulsive disorder: a step toward more effective treatments? In Goodman et al., 551–571.

Bobon D, Breulet M, Gerard-Vandenhove M-A, Guito-Goffioul F, Plomteux G, Satre-Hernandez M, Schratzer M, Troisfontaines B, von Frenckell R, Wachtel H. (1988). Is phosphodiesterase inhibition a new mechanism of antidepressant action? Eur Arch Psychiatr Neurol Sci, 238, 2–6.

Boguski MS. (2002). The mouse that roared. Nature, 420, 515–516.

Boolell M, Allen MJ, Ballard SA, Gepi-Attee S, Muirhead GJ, Naylor AM, Osterloh IH, Gingell C. (1996). Sildenafil: an orally active type 5 cyclic GMP-specific phosphodiesterase inhibitor for the treatment of penile erectile dysfunction. Int J Impot Res, 8, 47–52.

Borchelt DR, Ratovitski T, van Lare J, Lee MK, Gonzales V, Jenkins NA, Copeland NG, Price DL, Sisodia SS. (1997). Accelerated amyloid deposition in the brains of transgenic mice coexpressing mutant presenilin 1 and amyloid precursor proteins. Neuron, 19, 939–945.

Bradley C. (1937). The behavior of children receiving benzedrine. Am J Psychiatry, 94, 577–585.

Bradley C, Bowen M. (1941). Amphetamine (benzedrine) therapy of children's behavior disorders. Am J Orthopsychiatry, 11, 92–103.

Brambilla P, Barale F, Soares JC. (2001). Perspectives on the use of anticonvulsants in the treatment of bipolar disorder. Int J Neuropsychopharamcology, 4, 421–446.

Brix TH, Kyvik KO, Christensen K, Hegedus L. (2001). Evidence for a major role of heredity in Graves' disease: a population based study of two Danish twin cohorts. J Clin Endoc Metab, 86, 930–934.

Brown SL, Svrakic DM, Przybeck TR, Cloninger CR. (1992). The relationship of personality to mood and anxiety states: a dimensional approach. J Psychiatr Res, 26, 197–211.

Brown WA. (1998). The placebo effect. Scientific American, January, 90–95.

Bunney W, Davis J. (1965). Norepinephrine in depressive reactions. Arch Gen Psychiatry, 13, 483–494.

Burns DD. (1999). *The Feeling Good Handbook*. New York, Plume.

Bymaster FP, Dreshfield-Ahmad LJ, Threlkeld PG, Shaw JL , Thompson L, Nelson DL, Hemrick-Luecke SK, Wong DT. (2001). Comparative affinity of duloxetine and venlafaxine for serotonin and norepinephrine transporters in vitro and in vivo, human serotonin receptor subtypes, and other neuronal receptors. Neuropsychopharmacology, 25, 871–880.

Cade J. (1949). Lithium salts in the treatment of psychotic excitement. Medical Journal of Australia, 36, 349–352.

Carlsson A, Waters N, Waters S. Carlsson ML. (2000). Network interactions in schizophrenia—therapeutic implications. Brain Research Reviews, 31, 342–349.

Chen KK, Kao CH. (1926). Ephedrine and pseudoephedrine, their isolation, constitution, isomerism, properties, derivatives and synthesis. J Am Pharm Assoc, 15, 625–639.

Chemelli RM, Willie JT, Sinton CM, Elmquist JK, Scammell T, Lee C, Richardson JA, Williams SC, Xiong Y, Kisanuki Y, Fitch TE, Nakazato M, Hammer RE, Saper CB, Yanagisawa M. (1999). Narcolepsy in orexin knockout mice: molecular genetics of sleep regulation. Cell, 98, 437–451.

Conley RR, Taminga CA, Kelly DL, Richardson CM. (1999). Treatment-resistant schizophrenia patients respond to clozapine after olanzapine non-response. Biol Psychiatry, 46, 73–77.

Contarino A, Dellu F, Koob GF, Smith GW, Lee KF, Vale W, Gold LH. (1999). Reduced anxiety-like and cognitive performance in mice lacking the corticotropin-releasing factor receptor 1. Brain Research, 835, 1–9.

Corey-Bloom J, Anand R, Veach J. (1998). A randomized trial evaluating the efficacy and safety of ENA 713 (rivastigmine tartrate), a new acetylcholinesterase inhibitor in patients with mild to moderately severe Alzheimer's disease. Int J Geriatr Psychopharmacol, 1, 55–65.

Costa E, Guidotti A, Mao CC. (1975). Evidence for the involvement of GABA in the action of benzodiazepines: studies on rat cerebellum. Adv Biochem Psychopharmacol, 14, 113–130.

Cottle M. (1999). Selling shyness. The New Republic, Aug 2, 24–29.

Cravchik A, Goldman D. (2000). Neurochemical individuality: genetic diversity among human dopamine and serotonin receptors and transporters. Arch Gen Psychiatry, 57, 1105–1114.

Creese I, Burt DR, Snyder SH. (1976). Dopamine receptor binding predicts clinical and pharmacological potencies of antischizophrenic drugs. Science, 192, 481–483.

Davis KL, Charney D, Coyle JT, Nemeroff C (Eds.). (2002). *Neuropsychopharmacology: The Fifth Generation of Progress*. New York, Lippincott Williams and Wilkins.

DeFries JC, Gervais MC, Thomas EA. (1978). Response to 30 generations of open-field activity in laboratory mice. Behavior Genetics, 8, 3–13.

Delay J, Deniker P, Harl JM. (1952). Utilisation en thérapeutique psychiatrique d'une phénothiazine d'action centrale elective. Annales Medico-Psychologiques, 110, 112–131.

Dewachter I, Moechars D, van Dorpe J, Van den Haute C, Spittaels K, Van Leuven F. (2001). Modeling Alzheimer's disease in multiple transgenic mice. Biochem Soc Symp, 67, 203–210.

Duman RS. (1998). Novel therapeutic approaches beyond the serotonin receptor. Biol Psychiatry, 44, 324–335.

Duman RS, Heninger GR, Nestler EJ. (1997). A molecular and cellular hypothesis of depression. Arch Gen Psychiatry, 54, 597–606.

Dyke HJ, Montana JG. (2002). Update on the therapeutic potential of PDE4 inhibitors. Expert Opin Investig Drugs, 11, 1–13.

Egan MF, Goldberg TE, Kolachana BS, Callicott JH, Mazzanati

CM, Straub RE, Goldman D, Weinberger DR. (2001). Effect of COMT val 108/158 met genotype on frontal lobe function and risk for schizophrenia. PNAS, 98, 6917–6922.

Esler WP, Wolfe MS. (2001) A portrait of Alzheimer secretases— new features and familiar faces. Science, 293, 1449–1454.

Faraone SJ, Biederman J, Spencer T, Wilens T, Seidman LJ, Mick E, Doyle AE. (2000). Attention deficit/hyperactivity disorder in adults: an overview. Biol Psychiatry, 48, 9–20.

Flint J, Corley R, DeFries JC, Fulker, David W, Gray JA, Miler S, Collins AC. (1995). A simple genetic basis for a complex psychological trait in laboratory mice. Science, 269, 1432–1435.

Frank JD, Frank JB. (1991). *Persuasion and Healing: A Comparative Study of Psychotherapy*. Baltimore, Johns Hopkins University Press.

Freud S. (1917) A general introduction to psychoanalysis. Authorized English translation of the revised edition by Joan Riviere. New York, Washington Square Press.

Fromm MF, Kroemer HK, Eichelbaum M. (1997). Impact of P450 genetic polymorphism on the first-pass extraction of cardiovascular and neuroactive drugs. Advanced Drug Delivery Reviews, 27, 171–199.

Fujimaki K, Morinobu S, Duman RS. (2000). Administration of a cAMP phosphodiesterase 4 inhibitor enhances antidepressant-induction of BDNF mRNA in rat hippocampus. Neuropsychopharmacology, 22, 42–51.

Garver DL, Steinberg JL, McDermott BE, Yao JK, Ramberg JE, Lewis S, Kingsbury SJ. (1997). Etiologic heterogeneity of the psychoses: is there a dopamine psychosis? Neuropsychopharmacology, 16, 191–201.

Gladwell M. (1999). Running from Ritalin. The New Yorker, Feb 15, 80–84.

Glenner GG, Wong CW. (1984). Alzheimer's disease: initial report of the purification and characterization of a novel cerebrovascular amyloid protein. Biochem Biophys Research Com, 120, 885–890.

Goldberg D. (2000). Plato versus Aristotle: categorical and dimensional models for common mental disorders. Compr Psychiatry, 41 (2 Suppl 1), 8–13.

Goldstein DJ, Mallincrodt C, Lu Y, Demitrack MA. (2002). Duloxetine in the treatment of major depressive disorder. J Clin Psychiatry, 63, 225–231.

Goodman WK, Price LH, Delgado Pl, Palumbo J, Krystal JH, Nagy LM, Rasmussen JA, Heninger GR, Charney DJ. (1990). Specificity of serotonergic reuptake inhibitors in the treatment of obsessive-compulsive disorders: a comparison of fluvoxamine and desipramine. Arch Gen Psychiatry, 47, 577–585.

Goodman WK, Rudorfer MV, Maser JD (Eds.). (2000). *Obsessive-Compulsive Disorder: Contemporary Issues in Treatment*. Mahwah, N.J., Lawrence Erlbaum Associates.

Gorman CA, Garrity JA, Fatourechi V, Bahn RS, Stafford SL, Earle JD, Forbes GS, Bergstralh EJ, Offord KP, Rademacher DM, Stanley SM, Bartley JB. (2001). A prospective, randomized, double-blind, placebo-controlled study of orbital radiotherapy for Graves' ophthalmopathy. Ophthalmology, 108, 1523–1534.

Graves RJ. (1835). Palpitation of the heart with enlargement of the thyroid gland. London Medical and Surgical Journal (Renshaw's), 7, 516–517.

Graybiel AM, Rauch SL. (2000). Toward a neurobiology of obsessive-compulsive disorder. Neuron, 28, 343–347.

Graybiel AM, Saka E. (2002). A genetic basis for obsessive grooming. Neuron, 33, 1–2.

Greenblatt DJ, Shader RJ. (1971). Meprobamate: a study of irrational drug use. Am J Psychiatry, 127, 1297–1303.

Greer JM, Capecchi MR. (2002). Hoxb8 is required for normal grooming behavior in mice. Neuron, 33, 23–34.

Griffin LD, Mellon SH. (1999). Selective serotonin reuptake inhibitors directly alter activity of neurosteroidogenic enzymes. PNAS, 96, 13512–13517.

Gross C, Zhuang X, Stark K, Ramboz S, Oosting R, Kirby L, Santarelli L, Beck S, Hen R. (2002). Serotonin 1A receptor acts during development to establish normal anxiety-like behavior in the adult. Nature, 416, 396–400.

Guidotti A, Costa E. (1998). Can the antidysphoric and anxilolytic profiles of selective serotonin reuptake inhibitors be related to their ability to increase brain allopregnanolone availability? Biol Psychiatry, 44, 865–873.

Haefely W, Kulcsar A, Mohler H, Pieri L, Pole P, Schaffner R. (1975). Possible involvement of GABA in the central actions of benzodiazepines. Adv Biochem Psychopharmacol, 14, 131–151.

Hallowell EM, Ratey JJ. (1994). *Driven to Distraction: Recognizing and Coping with Attention Deficit Disorder from Childhood Through Adulthood.* New York, Touchstone.

Hamilton M. (1960). A rating scale for depression. Journal of Neurology, Neurosurgery and Psychiatry, 23, 56–62.

Hardy J, Selkoe DJ. (2002). The amyloid hypothesis of Alzheimer's disease: progress and problems on the road to therapeutics. Science, 297, 353–356.

Haskins JT, Moyer JA, Muth EA, Sigg EB. (1985). DMI, Wy-45,030, Wy-45,881, and ciramadol inhibit locus coeruleus neuronal activity. Eur J Pharmacol, 24, 139–146.

Hawson J, Luescher UA, Parent JM, Conlon PD, Ball RO. (1998). Efficacy of clomipramine in the treatment of canine compulsive disorder. J Am Vet Med Assoc, 213, 1760–1766.

Healy D. (1997). *The Antidepressant Era.* Cambridge, Mass., Harvard University Press.

Healy D. (2002). *The Creation of Psychopharmacology.* Cambridge, Mass., Harvard University Press.

Heresco-Levy U. (2000). N-Methyl-D-aspartate (NMDA) receptor-based treatment approaches in schizophrenia: the first decade. Int J Neuropsychopharmacol, 3, 243–258.

Higuchi S, Matsushita S, Muramatsu T, Murayama M, Hayashida M. (1996). Alcohol and aldehyde dehydrogenase genotypes and drinking behavior in Japanese. Alcoholism: Clinical and Experimental Research, 20, 493–497.

Hodge CN, Aldrich PE, Wasserman ZR, Fernandez CH, Nemeth GA, Arvanitis A, Cheeseman RS, Chorvat RJ, Ciganek E, Christos TE, Gilligan PJ, Krenitsky P, Scholfield E, Strucely P. (1999). Corticotropin-releasing hormone receptor antagonists: framework design and synthesis guided by ligand conformational studies. J Med Chem, 42, 819–832.

Hollander E, Allen A, Kwon J, Aronowitz B, Schmeidler J, Wong C, Simeon D. (1999). Clomipramine vs. desipramine crossover trial in body dysmorphic disorder: selective efficacy of a serotonin reuptake inhibitor in imagined ugliness. Arch Gen Psychiatry, 56, 1033–1039.

Hollander E, Liebowitz MR, Winchel R, Klumker A, Klein DF. (1989). Treatment of body-dysmorphic disorder with serotonin reuptake blockers. Am J Psychiatry, 146, 768–770.

Hollander E, Wong CM. (2000). Spectrum, boundary, and subtyping issues: implications for treatment-refractory obsessive-compulsive disorder. In Goodman et al., 3–22.

Hollister L. (1975). Drugs for emotional disorders. JAMA, 244, 942–947.

Hosogai N, Hamada K, Tomita M, Nagashima A, Takahashi T, Sekizawa T, Mizutani T, Urano Y, Kuroda A, Sawada K, Ozaki T, Seki J, Goto T. (2001). FR226807: a potent and selective phosphodiesterase type 5 inhibitor. Eur J Pharmacol, 428, 295–302.

Hyman SE, Nestler EJ. (1996). Initiation and adaptation: a paradigm for understanding psychotropic drug action. Am J Psychiatry, 153, 151–162.

Jensen PS, Mrazek D, Knapp PK, Steinberg L, Pfeffer C, Schowalter J, Shapiro T. (1997). Evolution and revolution in child psychiatry: ADHD as a disorder of adaptation. J Am Acad Child Adolesc Psychiatry, 36, 1672–1679.

Joyce PR, Mulder RT, Cloninger CR. (1994). Temperament predicts clomipramine and desipramine response in major depression. J Affect Disord, 30, 35–46.

Khan A, Leventhal RM, Khan SR, Brown WA. (2002). Severity of depression and response to antidepressants and placebo: an analysis of the Food and Drug Administration database. J Clin Psychopharmacol, 22, 40–45.

Kane J, Honigfeld G, Singer J, Meltzer H, and the Clozaril Collaborative Study Group. (1988). Clozapine for the treatment-resistant schizophrenic. Arch Gen Psychiatry, 45, 789–796.

Kanigel R. (1986). *Apprentice to Genius: The Making of a Scientific Dynasty*. New York, Macmillan.

Kaufman J, Charney D. (2000). Comorbidity of mood and anxiety disorders. Depress Anxiety, 12 Suppl 1, 69–76.

Kessler RC, McGonagle KA, Zhao S, Nelson CB, Hughes M, Eshleman S, Wittchen HU, Kendler KS. (1994). Lifetime and 12-month prevalence of DSM-III-R psychiatric disorders in the United States. Results from the National Comorbidity Survey. Arch Gen Psychiatry, 51, 8–19.

Kikuchi T, Tottori K, Uwahodo Y, Hirose T, Miwa T, Oshiro Y, Morita S. (1995). 7-(4-[4-(2,3-Dichlorophenyl)-1-piperazinyl]butyloxy)-3,4-dihydro-2(1H)-quinolinone (OPC-14597), a new putative antipsychotic drug with both presynaptic dopamine autoreceptor agonistic activity and postsynaptic D_2 receptor antagonistic activity. J Pharmacol Exp Ther, 274, 329–336.

Klein DF, Fink M. (1962). Psychiatric reaction patterns to imipramine. Am J Psychiatry, 119, 432–438.

Klerman GL. (1990). The psychiatric patient's right to effective treatment: implications of Osheroff vs. Chestnut Lodge. Am J Psychiatry, 147, 409–418.

Kozak MJ, Liebowitz MR, Foa EB. (2000). Cognitive behavior therapy and pharmacotherapy for obsessive-compulsive disorder: the NIMH-sponsored collaborative study. In Goodman et al., 501–530.

Kramer MS, Cutler N, Feighner J, Shrivastava R, Carman J, Sramek JJ, Reines SA, Liu G, Snavely D, Wyatt-Knowles E, Hale JJ, Mills SG, MacCoss M, Swain CJ, Harrison T, Hill RG, Hefti F, Scolnick EM, Cascieri MA, Chicchi GG, Sadowski S, Williams AR, Hewson L, Smith D, Carlson EJ, Hargreaves RJ, Rupniak NM. (1998). Distinct mechanism for antidepressant activity by blockade of central substance P receptors. Science, 281, 1640–1645.

Kramer PD. (1993). *Listening to Prozac*. New York, Viking.

Krystal JH, D'Souza DC. (1998). D-serine and the therapeutic challenge posed by the N-methyl-D-aspartate antagonist model of schizophrenia. Biol Psychiatry, 44, 1075–1076.

Kuhn R. (1958). The treatment of depressive states with G22355 (imipramine hydrochloride). Am J Psychiatry, 115, 459–464.

Lambert PA, Cavas G, Borselli S, Carbel S. (1966). Action neuropsychotrope d'un nouvel anti-epileptique, le depamide. Annales Medico-Psychologiques (Paris), 1, 707–710.

Langreth R. (2002). Betting on the brain. Forbes, Jan 7, 56–59.

Leary AC, Grealy G, Higgins TM, Buckley N, Barry DG, Murphy D, Ferriss JB. (1999). Long-term outcomes of treatment of hyperthyroidism in Ireland. Irish J of Medical Sci, 168, 47–52.

Leake C. (1958). *The Amphetamines*. Springfield, CC Thomas.

Lendon CL, Ashall F, Goate AM. (1997). Exploring the etiology of Alzheimer disease using molecular genetics. JAMA, 277, 825–831.

Lieberman J, Jody D, Geisler S, Alvir J, Loebel A, Szymanski S, Woerner M, Borenstein M. (1993). Time course and biologic correlates of treatment response in first-episode schizophrenia. Arch Gen Psychiatry, 50, 369–376.

Lin L, Faraco J, Li R, Kadotani H, Rogers W, Lin X, Qiu X, de Jong P, Nishino S, Mignot E. (1999). The sleep disorder canine narcolepsy is caused by a mutation in the hypocretin (orexin) receptor 2 gene. Cell, 98, 365–376.

Lin L, Hungs M, Mignot E. (2001). Narcolepsy and the HLA region. J Neuroimmunol, 117, 9–20.

Litman RE, Su TP, Potter WZ, Hong WW, Pickar D. (1996). Idazoxan and response to typical neuroleptics in treatment-resistant schizophrenia. Comparison with the atypical neuroleptic, clozapine. Br J Psychiatry, 168, 571–579.

Low K, Crestani F, Keist R, Benke D, Brunig I, Benson JA, Fritschy J-M, Rulicke T, Bluethmann H, Mohler H, Rudolph U. (2000). Molecular and neuronal substrate for the selective attenuation of anxiety. Science, 290, 131–134.

Mahgoub A, Idle JR, Dring LG, Lancaster R, Smith RL. (1977). Polymorphic hydroxylation of debrisoquine in man. Lancet, 2 (8038), 584–586.

Manji HK, Drevets WC, Charney DS. (2001). The cellular neurobiology of depression. Nature Medicine, 7, 541–547.

Manji HK, Zarate CA. (2002). Molecular and cellular mechanisms underlying mood stabilization in bipolar disorder: implications for the development of improved therapeutics. Mol Psychiatry, 7 Suppl 1, S1–7.

Maurer K, Volk S, Gerbaldo H. (1997). Auguste D. and Alzheimer's disease. Lancet, 349, 1546–1549.

McCarthy JR, Heinrichs SC, Grigoriadis DE. (1999). Recent advances with the CRF1 receptor: design of small molecule inhibitors, receptor subtypes and clinical indications. Curr Pharm Des, 5, 289–315.

McKernan RM, Rosahl TW, Reynolds DS, Sur C, Wafford KA, Attack JR, Farrar S, Myers J, Cook G, Ferris P, Garrett L, Bristow G, Marshall A, Macaulay N, Brown N, Howell O, Moore KW, Carling RW, Street LJ, Castro JL, Ragan CI, Dawson GR, Whiting PJ. (2000). Sedative but not anxiolytic properties of benzodiazepines are mediated by the GABA-A receptor alpha-1 type. Nature Neuroscience, 3, 587–592.

Meltzer HY, Stahl SM. (1976). The dopamine hypothesis of schizophrenia: a review. Schizophrenia Bulletin, 2, 19–76.

Moseley JB, O'Malley K, Petersen NJ, Menke TJ, Brody BA, Kuykendall DH, Hollingsworth JC, Ashton CM, Wray NP. (2002). A controlled trial of arthroscopic surgery for osteoarthritis of the knee. N Engl J Medicine, 347, 81–88.

MTA Cooperative Group (1999). A 14-month randomized clinical trial of treatment strategies for attention-deficit/hyperactivity disorder. Arch Gen Psychiatry, 56, 1073–1096.

Mundo E, Walker M, Cate T, Macciardi F, Kennedy JL. (2001). The role of serotonin transporter protein gene in antidepressant-induced mania in bipolar disorder: preliminary findings. Arch Gen Psychiatry, 58, 539–544.

Murphy DL, Li Q, Engel S, Wicherns C, Andrews A, Lesch K-P, Uhl G. (2001). Genetic perspectives on the serotonin transporter. Brain Research Bulletin, 56, 487–494.

Myerson A. (1936). Effect of benzedrine sulfate on mood and fatigue in normal and neurotic persons. Arch Neurol Psychiat, 36, 816–822.

National Institute of Mental Health, Psychopharmacology Service Center Collaborative Study Group (1964). Phenothiazine treatment in acute schizophrenia. Effectiveness. Arch Gen Psychiatry, 10, 246–261.

Nebert DW. (1997). Polymorphisms in drug-metabolizing enzymes: what is their clinical relevance and why do they exist? Am J Hum Genet, 60, 265–271.

Nestler EJ, Hyman SE, Malenka RC. (2001). *Molecular Neuropharmacology: A Foundation for Clinical Neuroscience.* New York, McGraw-Hill.

Okuma T, Kishimoto A, Inoue K, Matsumoto H, Ogura A. (1973). Anti-manic and prophylactic effects of carbamazepine (Tegretol) on manic-depressive psychosis. A preliminary report. Folia Psychiatrica Neurologica Japonica, 27, 283–297.

Overall KL. (2000). Natural animal models of human psychiatric conditions: assessment of mechanism and validity. Progress Neuro-Psychopharmacological and Biological Psychiatry, 24, 727–776.

Ozdemir V, Basile VS, Masellis M, Kennedy JL. (2001). Pharmacogenetic assessment of antipsychotic-induced movement disorders: contribution of the dopamine D3 receptor and cytochrome P450 1A2 genes. J Biochem Biophys Methods, 47, 151–157.

Pauls DL, Alsobrook JP, Goodman WK, Rasmussen S, Leckman JF. (1995). A family study of obsessive-compulsive disorder. Am J Psychiatry, 152, 76–84.

Phillips KA. (1986). *The Broken Mirror: Understanding and Treating Body Dysmorphic Disorder.* New York, Oxford University Press.

Phillips KA, Albertini RS, Rasmussen SA. (2002). A randomized placebo-controlled trial of fluoxetine in body dysmorphic disorder. Arch Gen Psychiatry, 59, 381–388.

Phillips KA, McElroy SL, Hudson JI, Pope HG Jr. (1995). Body dysmorphic disorder: an obsessive-compulsive spectrum disorder, a form of affective spectrum disorder, or both? J Clin Psychiatry, 56 Suppl 4, 41–51.

Pincus HA, Zarin DA, Tanielian TL, Johnson JL, West JC, Pettit AR, Marcus SC, Kessler RC, McIntyre JS. (1999). Psychiatric patients and treatments in 1997: findings from the American Psychiatric Practice Research Network. Arch Gen Psychiatry, 56, 441–449.

Prinzmetal M, Bloomberg W. (1935). The use of benzedrine for the treatment of narcolepsy. JAMA, 105, 2051–2054.

Quitkin FM, Rabkin JG, Gerald J, Davis JM, Klein DF. (2000). Validity of clinical trials of antidepressants. Am J Psychiatry, 157, 327–337.

Quitkin FM, Rabkin JG, Gross D, Stewart JW. (1984). Identification of true drug response to antidepressants: use of pattern analysis. Arch Gen Psychiatry, 41, 782–786.

Quitkin FM, Stewart JW, McGrath PJ, Nune E, Ocepek-Welikson K, Tricamo E, Rabkin JG, Ross D, Klein DF. (1993). Loss of drug effects during continuation therapy. Am J Psychiatry, 150, 562–565.

Raeymaekers P, Van Broeckhoven C. (1998). Comment—Genes and temperament, a shortcut for unravelling the genetics of psychopathology? Int J Neuropsychopharmacol, 1, 169–171.

Rafal S, Tsuang MT, Carpenter WT Jr. (1999). A dilemma born of progress: switching from clozapine to a newer antipsychotic. Am J Psychiatry, 156, 1086–1090.

Rapoport JL, Buchsbaum M S, Weingartner H, Zahn TP, Ludlow C, Mikkelsen EJ. (1980). Dextroamphetamine: Its cognitive and behavioral effects in normal and hyperactive boys and normal men. Arch Gen Psychiatry, 37, 933–943.

Report of the National Institute of Mental Health's Genetics Workgroup (1999). Biol Psychiatry, 45, 559–602.

Reus VI, Wolkowitz OM. (2001). Antiglucocorticoid drugs in the treatment of depression. Expert Opin Investig Drugs, 10, 1789–1796.

Rho JM, Donevan SD, Rogawski MA. (1997). Barbiturate-like actions of the propanediol dicarbamates felbamate and meprobamate, J Pharmacol Exper Ther, 280, 1383–1391.

Risch N. (2000). Searching for genetic determinants in the new millenium. Nature, 405, 847–856.

Rockwood K, Mintzer J, Truyen L, Wessel T, Wilkinson D. (2001). Effects of a flexible galantamine dose in Alzheimer's disease: a randomized, controlled trial. J Neurol Neurosurg Psychiatry, 71, 589–595.

Rogers SL, Farlow MR, Doody RS, Mohs R, Friedhoff LT, and the Donezepil Study Group. (1998). A 24-week, double-blind, placebo-controlled trial of donezepil in patients with Alzheimer's disease. Neurology, 50, 136–145.

Roses AD. (2000). Pharmacogenetics and the practice of medicine. Nature, 405, 857–865.

Rudolph U, Crestani F, Benke D, Brunig I, Benson JA, Fritschy J-M, Martin JR, Bluethmann H, Mohler H. (1999). Benzodiazepine actions mediated by specific gamma aminobutyric acid-A receptor subtypes. Nature, 401, 796–800.

Rupniak NM, Kramer MS. (1999). Discovery of the antidepressant and anti-emetic efficacy of substance P receptor (NK1) antagonists. Trends Pharmacol Sci, 20, 485–490.

Sachse C, Brockmoller J, Bauer S, Roots I. (1997). Cytochrome P450 2D6 variants in a Caucasian population: allele frequencies and phenotypic consequences. Am J Hum Genet, 60, 284–295.

Saxena S, Brody AL, Schwartz JM, Baxter LR. (1998). Neuroimaging and frontal-subcortical circuitry in obsessive-compulsive disorder. Br J Psychiatry Suppl, 35, 26–37.

Scahill L, Chappell PB, Kim YS, Scultz RT, Katsovich L, Shepherd E, Arnste AFT, Cohen DJ, Leckman JF. (2001). A placebo-controlled study of guanfacine in the treatment of children with tic disorders and attention deficit hyperactivity disorder. Am J Psychiatry, 158, 1067–1074.

Schildkraut JJ. (1965). The catecholamine hypothesis of affective disorders: a review of supporting evidence. Am J Psychiatry, 122, 509–522.

Schildkraut JJ, Kety SS. (1967). Biogenic amines and emotion. Science, 156, 21–30.

Schrof JM, Schultz S. (1999.) Social Anxiety. U.S. News and World Report, Jun 21, 50–57.

Seeman P, Lee T, Chai-Wong M, Wong K. (1976). Antipsychotic drug doses and neuroleptic/dopamine receptors. Nature, 261, 717–718.

Selkoe DJ. (1999). Translating cell biology into therapeutic advances in Alzheimer's disease. Nature, 399 (6738 suppl.), A23–A31.

Selling LS. (1955). Clinical study of a new tranquilizing drug. Use of

Miltown (2-methyl-2-n-propyl-1,3 propanediol dicarbamate). JAMA, 157, 1594–1596.

Shorter E. (1997). *A History of Psychiatry.* New York, John Wiley.

Siegel JM, Moore R, Thannickal T, Nienhuis R. (2001). A brief history of hypocretin/orexin and narcolepsy. Neuropsychopharmacology, 25, S5, S14–S20.

Simons M, Keller P, Dichgans J, Schutz JB. (2001). Cholesterol and Alzheimer's disease: is there a link? Neurology, 57, 1089–1093.

Smith MC. (1991). *A Social History of the Minor Tranquilizers: The Quest for Small Comfort in the Age of Anxiety.* Binghamton, N.Y., Haworth Press.

Snyder SH. (1986). *Drugs and the Brain.* New York, Scientific American Library.

Snyder SH, Ferris CD. (2000). Novel neurotransmitters and their neuropsychiatric relevance. Am J Psychiatry, 157, 1738–1751.

Solomon A. (1998). Anatomy of melancholy. The New Yorker, Jan 12.

Solomon A. (2001). *The Noonday Demon: An Atlas of Depression.* New York, Scribner.

Stahl SM. (2000). *Essential Psychopharmacology: Neuroscientific Basis and Practical Applications*, second edition. Cambridge, Cambridge University Press.

Stahl SM. (2001). Dopamine system stabilizers, aripiprazole, and the next generation of antipsychotics, part 1, "Goldilocks" actions at dopamine receptors. J Clin Psychiatry, 62, 841–842.

Stefansson H, Sigurdsson E, Steinthorsdottir V, Bjornsdottir S, Sigmundsson T, Ghosh S, Brynjolfsson J, Gunnarsdottir S, Ivarsson O, Chou TT, Hjaltason O, Birgisdottir B, Jonsson H, Gudnadottir VG, Gudmundsdottir E, Bjornsson A, Ingvarsson B, Ingason A, Sigfusson S, Hardardottir H, Harvey RP, Lai D, Zhou M, Brunner D, Mutel V, Gonzalo A, Lemke G, Sainz J, Johannesson G, Andresson T, Gudbjartsson D, Manolescu A, Frigge ML, Gurney ME, Kong A, Gulcher JR, Petursson H, Stefansson K. (2002). Neuregulin 1 and susceptibility to schizophrenia. Am J Hum Genet, 71, 877–892.

Stein MB, Liebowitz MR, Lydiard RB, Pitts CD, Bushnell W, Gergel I. (1998). Paroxetine treatment of generalized social phobia (social anxiety disorder}: a randomized controlled trial. JAMA, 280, 708–713.

Stenszky V, Kozma L, Balazs C, Rochlitz S, Bear JC, Farid NR. (1985). The genetics of Graves' disease: HLA and disease susceptibility. Journal of Clinical Endocrinology and Metabolism, 61, 735–740.

Sternbach L. (1979). The benzodiazepine story. Journal of Medicinal Chemistry, 22, 1–7.

Stewart WF, Kawas C, Corrada M, Metter EJ. (1997). Risk of Alzheimer disease and duration of NSAID use. Neurology, 48, 626–632.

Stipp D. (2000). A DNA tragedy. Fortune, Oct 30.

Straub RE, Jiang Y, MacLean CJ, Ma Y, Webb BT, Myakishev MV, Harris-Kerr C, Wormley B, Sadek H, Kadambi B, Cesare AJ, Gibberman A, Wang X, O'Neill FA, Walsh D, Kendler KS. (2002). Genetic variation in the 6p22.3 gene DTNBP1, the human ortholog of the mouse dysbindin gene, is associated with schizophrenia. Am J Hum Genet, 71, 337–348.

Strittmater WJ, Roses AD. (1996). Apolipoprotein E and Alzheimer disease. Annual Review of Neuroscience, 19, 53–77.

Swazey, Judith P. (1974). *Chlorpromazine in Psychiatry: A Study in Therapeutic Innovation.* Cambridge, Mass., MIT Press.

Tamminga CA. (2002). Partial dopamine agonists in the treatment of psychosis. J Neural Transm, 109, 411–20.

Tanzi RE, Parson AB. (2000). *Decoding Darkness: The Search for the Genetic Causes of Alzheimer's Disease.* New York, Perseus.

Tecott LH, Wehner JM. (2001). Mouse molecular genetic technologies: promise for psychiatric research. Arch Gen Psychiatry, 58, 995–1004.

Thase ME. (1999). How should efficacy be evaluated in randomized clinical trials of treatments for depression? J Clin Psychiatry, 60 (suppl 4), 23–31.

Thase ME, Entsuah AR, Rudolph RL. (2001). Remission rates during treatment with venlafaxine or selective serotonin reuptake inhibitors. Br J Psychiatry, 178, 234–41.

Thase ME, Fava M, Halbreich U, Kocsis JH, Koran L, Davidson J,

Rosenbaum J, Harrison W. (1996). A placebo-controlled, randomized clinical trial comparing sertraline and imipramine for the treatment of dysthymia. Arch Gen Psychiatry, 53, 777–784.

Thompson A. (2001). Paxil maker held liable in murder/suicide. Lawyer's Weekly, Jul 9.

Toufexis A. (1993). The personality pill. Time, Oct 11, 61–62.

Tsai G, Yang P, Chung LC, Lange N, Coyle JT. (1998). D-serine added to antipsychotics for the treatment of schizophrenia. Biol Psychiatry, 44, 1081–1089.

Turri MG, Henderson ND, Defries JC, Flint J. (2001). Quantitative trait locus mapping in laboratory mice derived from a replicated selection experiment for open-field activity. Genetics, 158, 1217–1226.

Uzunova V, Sheline Y, Davis JM, Rasmusson A, Uzunov DP, Cost E, Guidotti A. (1998). Increase in the cerebrospinal fluid content of neurosteroids in patients with unipolar major depression who are receiving fluoxetine or fluvoxamine. PNAS, 95, 3239–3244.

van Leuven F. (2000). Single and multiple transgenic mice as models for Alzheimer's disease. Prog Neurobiol, 61, 305–312.

Viaterna MH, King DP, Chang AM, Kornhauser JM, Lowrey PL, McDonald JD, Dove WF, Pinto LH, Turek FW, Takahashi JS. (1994). Mutagenesis and mapping of a mouse gene, Clock, essential for circadian behavior. Science, 264, 719–725.

Vogel G. (2001). How the body's "garbage disposal" may inactivate drugs. Science, 291, 35–37.

Wang Z, Gorski JC, Hamman MA, Huang SM, Lesko LJ, Hall SD. (2001). The effects of St. John's wort (Hypericum perforatum) on human cytochrome P450 activity. Clin Pharmacol Ther, 70, 317–326.

Weetman AP. (2001). Determinants of autoimmune thyroid disease. Nature Immunology, 2, 769–770.

Wilens TE, Biederman J, Prince J, Spencer TJ, Faraone SV, Warburton R, Schleifer D, Harding M, Linehan C, Geller D. (1996). Six-week, double-blind, placebo-controlled study of desipramine for adult attention deficit hyperactivity disorder. Am J Psychiatry, 153, 1147–1153.

Wong DT, Bymaster FP, Engleman EE. (1995). Prozac (fluoxetine, Lilly 110140), the first selective serotonin uptake inhibitor and an antidepressant drug: twenty years since its first publication. Life Sciences, 57, 411–441.

Wynchank DR, Berk M. (1998). Fluoxetine treatment of acral lick dermatitis in dogs: a placebo-controlled randomized double blind trial. Depression and Anxiety, 8, 21–23.

Zanardi R, Seretti A, Rossini D, Franchini L, Cusin C, Lattuada E, Dotoli D, Smeraldi E. (2001). Factors affecting fluvoxamine antidepressant activity: influence of pindolol and 5-HTTLPR in delusional and non-delusional depression. Biol Psychiatry, 50, 323–330.

Zobel AW, Nickel T, Kunzel HE, Ackl N, Sonntag A, Ising M, Holsboer F. (2000). Effects of the high-affinity corticotropin-releasing hormone receptor 1 antagonist R121919 in major depression: the first 20 patients treated. J Psychiatr Res, 34, 171–181.

Acknowledgments

Many people helped me write this book. I am grateful to Ira Herskowitz and Cyra McFadden for detailed suggestions and to Corey Goodman, James Leckman, Julio Licinio, Alan Louie, Husseini Manji, Steven Paul, Victor Reus, Stephen Stahl, Larry Tecott and Owen Wolkowitz for valuable comments. I also thank Owen for inviting me to participate in his Psychopharmacology Clinic at Langley Porter Psychiatric Institute.

I learned a lot about clinical psychiatric research by serving, over the past few years, as Chair of the Board of Scientific Counselors of the National Institute of Mental Health. I also benefited from more than a decade of education in psychiatric drug development through my membership on the scientific advisory boards of DuPont-Merck Pharmaceuticals, Guilford Pharmaceuticals, and Renovis Inc., and by interacting with many excellent scientists at those companies.

I thank Kirk Jensen, my editor in the United States, and Michael Rodgers, my editor in the United Kingdom, for helpful recommendations. I thank Helen Mules for assistance with the manuscript. I thank Katinka Matson, my agent at Brockman Inc., for thoughtful advice.

I am especially grateful for the love and devotion of my wife, Louann Brizendine, and my daughters, Elizabeth and Jessica. Louann also enriched this book by sharing many insights from her inspired practice of psychiatry.

Index

chlorpromazine (*continued*)
 effect on neurotransmission,
  23, 25, 160
 features, 69
 metabolism, 131
 onset of action, 25, 46
cholesterol, 103–4
chromosomes, 105–6
Ciba (pharmaceutical com-
 pany), 71
circadian behavior, 177n (p118)
citalopram (Celexa), 43, 57,
 133, 171n (p58)
Clara's example, 3–16, 90–92,
 93, 153–55, 158–61
Claritin, 18
"clean" drugs, 142
clomipramine (Anafranil), 15,
 113, 159
clonazepam (Klonopin), 54,
 124
clozapine (Clozaril), 25–27,
 27–28, 167n (p28)
cognitive therapy, 4, 5, 16, 49,
 159
combinations of medications,
 154, 181n (p154)
complex gene disorders, 107
Comprehensive Drug Abuse
 Prevention and Control
 Act (1970), 70
COMT (catechol-o-methyl-
 transferase), 36, 106, 110
COMT gene variants, 106, 110
Concerta (methylphenidate),
 72, 143

corticotropin-releasing factor
 (CRF), 147
corticotropin-releasing hor-
 mone (CRH), 147–49
cortisol (hydrocortisone),
 147–48, 181n (p148)
Costa, Erminio, 54
Cottle, Michelle, 145
Council of Pharmacy and
 Chemistry of the American
 Medical Association, 63
CRF (corticotropin-releasing
 factor), 147
CRH-ACTH system, 147–48
CRH (corticotropin-releasing
 hormone), 147–49
cyclic AMP, 45, 149. 150
cyclic GMP, 45, 149, 150
Cymbalta (duloxetine), 143
cytochrome P-450 (CYP),
 130–33
CYP3A, 131–32
CYP3A4, 131
CYP2D6 (cytochrome P-450-
 2D6, or 2D6), 130, 131,
 133
CYP1A2, 138

D-serine, 30
Davis, John, 38
debrisoquine, 132–33, 134
DeCODE Genetics, 108–9
DeFries, John, 117
Delay, Jean, 19
delayed drugs, 46, 58, 68, 69
Deniker, Pierre, 19, 166n (p20)

117, 118, 120, 127, 133,
138–39, 156
genetic drift, 134
genetic engineering in mice,
119–20
genetic heterogeneity, 107–8
Grave's disease, 88–90
major depression, 105
multiple gene disorders, 107
mutations, 117, 118, 177n
(p118)
number of genes, 98, 176n
(p98)
obsessive-compulsive disorder, 105, 155–56
panic disorder, 105
schizophrenia, 90, 104,
106–10
susceptibility gene variants,
107
therapeutic applications, 92
Tourette's disorder, 155–56
Geodon (ziprasidone), 28, 143
Gladwell, Malcolm, 74–75
GlaxoSmithKline (pharmaceutical company), 135–36
glutamate, 22, 29–30
glycine, 30
Goate, Alison, 99
granulocytes, 25
Graves, Robert, 79, 82
Graves' disease, 81–92, 174n
(p87)
Greenblatt, David, 51–52
Greer, Joy, 156–57, 158, 160
guanfacine (Tenex), 75–76

Guillain-Barré syndrome, 40

Haefely, Willy, 54
Halcion (triazolam), 54
Haldol (haloperidol), 24, 26,
27–28, 30, 138
Hallowell, Edward, 74
haloperidol (Haldol), 24, 26,
27–28, 30, 138
Hamilton, Max, 42
Hamilton Depression Scale
(Ham-D), 42–43, 44, 142
Handbook of Psychiatry (Kraepelin), 96
Hardy, John, 99
herbal remedies, 131–32
heroin, 70
heterogeneity, genetic, 107–8
histamine, 23, 142
Hoffman–La Roche (pharma-
ceutical company), 34,
52–53, 125, 126
Hollander, Eric, 155
Hollister, Leo, 77, 79–80
Hoxb8 gene, 156–57, 158
Human Genome Project, 109
humans, testing drugs on, 41,
180n (p144)
hydrocortisone (cortisol),
147–48, 181n (p148)
hyperthyroidism, 83, 92. *See
also* Graves' disease
hypocretins (orexins), 121–22,
123
hypothalamus, 147, 148
hypothyroidism, 83, 92

ibuprofen, 104
Iceland, 108–9
"ideal psychotherapeutic drug,"
    79, 173n (p79)
imipramine (Tofranil)
    as antidepressant, 31–39, 57,
        79, 171n (p57)
    as "dirty" drug, 142
    discovery, 47
    effect on anxiety, 56–57
    effect on neurotransmission,
        37, 38, 160
    metabolism, 131
immediate drugs, 46, 58, 59,
    67, 68, 69
inbred strains of mice, 114–18
inhibitory signals, 23
insomnia, 32–33, 34, 72
iproniazid, 34–35, 37, 38

Jackson Laboratory, 114
Janssen (pharmaceutical com-
    pany), 24
Japanese, gene variants in, 134,
    137
Jensen, Peter, 72

Kendler, Kenneth, 109
Kety, Seymour, 17, 38
Klein, Donald, 56–57
Kline, Nathan, 21, 34
Klonopin (clonazepam), 54, 124
"knock-in" mice, 123–26
"knockout" mice, 120–23, 148,
    156–57
Koetschet, Pierre, 18

Kraepelin, Emil, 96
Kramer, Peter, 9
Kuhn, Roland, 31, 33–34,
    41–43, 56

L-838,417 (experimental drug),
    125
L-dopa, 22
Laborit, Henri, 18–19
lamotrigine (Lamictal), 169n
    (p46)
Largactil, 20. See also chlorpro-
    mazine
Leckman, James, 155–56
Librium (chlordiazepoxide),
    52–53, 54
light-dark box test, 116
limbic system, 26
linkage approach, 105–6
Listening to Prozac (Kramer),
    9
lithium carbonate, 46, 169n
    (p46)
London gene variant, 99–100,
    105, 108, 119
long-acting thyroid stimulator
    (LATS), 87, 174n (p87)
LSD, 70
Ludwig, B. J., 50
Luminal (phenobarbital),
    17–18, 49–52, 55, 123–24,
    172n (p68)
Luvox (fluvoxamine), 43, 57,
    137, 171n (p58)

ma huang, 63

magnetic resonance imaging (MRI), 110

major depression. *See* depression

male erectile disorder, 150

MAO inhibitors, 39, 57, 163, 171n (p57)

MAO (monoamine oxidase), 35–37

markers (genetic), 105–6

marketing drugs, 52, 143

Martha's example, 80–81, 84–85, 90–92, 93

medical disorders, relationship to mental disorders, 90–91, 175n (p90)

Mendel, Gregor, 99

Mendelian inheritance, 99

mental disorders, definition of, 90–91

meprobamate (Equanil), 50, 141–42

meprobamate (Miltown), 50–52, 53, 55

Merck (pharmaceutical company), 39, 125–26, 146–47, 180n (p144)

metabolism of drugs, 130–34

Metadate-CD (methylphenidate), 72

methimazole, 83–84, 85–86, 92

methylphenidate (Ritalin), 71–74, 74–76, 143, 164

me-too medication, 25

mice used in behavioral research, 114–27, 148,

156–57, 177n (p118)

Mignot, Emanuel, 121–22

Miltown (meprobamate), 50–52, 53, 55

mirtazapine (Remeron), 43

MK-869 (experimental drug), 146–47, 180n (p147)

modafinil (Provigil), 123

monoamine oxidase inhibitors. *See* MAO inhibitors

mood stabilizers, 46, 164

morphine, 146

Motrin (ibuprofen), 104

mouse genome project, 127

mouse behavioral tests, 116

multiple gene disorders (or complex gene disorders), 107

mutations in genes, 117, 118, 177n (p118)

Myerson, Abraham, 65

narcolepsy, 64, 71, 121–22

National Institute of Mental Health (NIMH), 72

National Institutes of Health (NIH), 36

Nature Reviews Drug Discovery, 158–59

nefazodone (Serzone), 43, 131

negative symptoms, 26, 27, 29

nerve terminals, 22

neuregulin-1, 109

Neurocrine Biosciences (pharmaceutical company), 149

neurokinins, 145–46

Neurontin (gabapentin), 169n (p46)
neuroticism, 137
neurotransmitters. *See also* acetylcholine; dopamine; GABA; glutamate; norepinephrine; serotonin
drugs' influence on, 35, 37, 43, 154, 160
explanation of, 22–25
receptors, 27, 29
The New Republic, 145
The New Yorker, 32–33, 43–44, 74–75
NMDA receptors, 29–30
A Noonday Demon (Solomon), 32
norepinephrine
amphetamine's effect on, 67, 69, 75–76
atomoxetine's effect on, 173n (p75)
augmentation of, 43, 154, 169n (p43)
COMT's effect on, 36
desipramine's effect on, 39, 173n (p75)
duloxetine's effect on, 143
GABA compared to, 54
MAO's effect on, 35
methylphenidate's effect on, 72, 75–76
Modafinil's effect on, 123
norepinephrine hypothesis of depression, 37–39, 57

norepinephrine transporter, 37
nortriptyline's effect on, 39
popular interest, 9
Prozac's effect on, 40
receptors, 44
venlafaxine's effect on, 142
nortriptyline, 39
nucleotides, 97
nucleus accumbens, 68

obesity, 28, 143
obsessive-compulsive disorder (OCD)
in animal psychiatry, 113–14
anxiety component, 49
body dysmorphic disorder and, 155, 182n (p155)
Clara's potential diagnosis of, 15
drug combinations for, 159
genetic component, 105, 156–57
OCD circuit, 155–57
OC spectrum disorders, 155–58
off-label uses of drugs, 7, 15, 143–44, 165n (p7)
olanzapine (Zyprexa), 28
one-size-fits all medications, 139
onset of action of drugs, 25, 45–46, 169n (p45), 169n (p46). *See also* delayed drugs, immediate drugs
open-field test, 116
orbito-frontal cortex, 155

orexins (hypocretins), 121–22, 123
Osheroff, Rafael, 136
Osler, William, 129
Otsuka Pharmaceuticals, 29
Overall, Karen, 113–14
overdoses, 54, 55, 129–30

pain, 145–46
panic disorder, 48, 56–58, 84, 105
paranoia, 69, 157
Parkinson, James, 21
Parkinson's-disease-like symptoms, 21, 23, 26, 27, 28
Parnate (tranylcypromine), 39
paroxetine (Paxil), 43, 57, 131, 135–36, 144–45, 171n (p58)
Parry, Caleb, 81
pathogenesis, 79–80, 82, 102, 111
pathophysiology, 82, 147
Pauls, David, 155–56
Paxil (paroxetine), 43, 57, 131, 135–36, 144–45, 171n (p58)
Pfizer (pharmaceutical company), 58, 126, 143, 144, 149–50
pharmaceutical companies, 102, 141, 144, 180n (p144). See also specific companies
pharmacodynamic differences, 134–38
pharmacokinetic differences, 134–36, 138
pharmacological effect vs. placebo effect, 11, 15

phases of drug testing, 180n (p144)
phencyclidine, 29
Phenergan (promethazine), 18, 21, 23
phenobarbital (Luminal), 17–18, 49–52, 55, 123–24, 172n (p68)
Phillips, Katherine, 155
phosphodiesterase inhibitors, 149, 150
phosphodiesterases, 149
physical disorders, relationship to mental disorders, 90–91, 175n (p90)
pituitary gland, 37, 86
placebo-controlled study, 11
placebo effect, 11, 12, 13–14, 41–43, 115, 166n (p12), 166n (p14), 168n (p43)
positive symptoms, 29
"Practice Guidelines for the Treatment of Panic Disorder" (American Psychiatric Association), 57–58
prefrontal cortex, 110
prescription, excessive, 56, 61
presenilin-1, 100–101, 106, 111, 119
presenilin-2, 100–101, 106, 108, 111
primates used in behavioral research, 113, 114
Prolixin (fluphenazine), 24, 26
promethazine (Phenergan), 18, 21, 23

ultrarapid metabolizers of
    drugs, 133
U.S. News & World Report, 144

Valium (diazepam), 53–54, 69,
    79, 85, 123–26, 133, 172n
    (p68)
valproate (valproic acid,
    Depakote), 46, 169n (p46)
venlafaxine (Effexor), 43, 142,
    143
Viagra (sildenafil), 149–51
Vickery, Andy, 135
viral infections, 89
Volga Germans, 108

Wallace Laboratories, 50
Wellbutrin (bupropion), 43, 154

withdrawal symptoms, 55–56
Wong, David, 40
World Health Organization, 32
Wyeth (pharmaceutical com-
    pany), 50–52, 141–42

Xanax (alprazolam), 54, 124, 131
Xanax XR, 54

Yanigasawa, Masashi, 121

Zelmid (zimelidine), 40–41
ziprasidone (Geodon), 28, 143
Zocor (simvastatin), 103–4
Zoloft (sertraline), 43, 57, 58,
    144, 171n (p58)
Zyban (bupropion), 154
Zyprexa (olanzapine), 28